George Wightwick

Hints to Young Architects,

Comprising Advice to those who, while yet at School, are Destined... Fifth Edition

George Wightwick

Hints to Young Architects,
Comprising Advice to those who, while yet at School, are Destined... Fifth Edition

ISBN/EAN: 9783337211158

Printed in Europe, USA, Canada, Australia, Japan

Cover: Foto ©berggeist007 / pixelio.de

More available books at **www.hansebooks.com**

HINTS TO YOUNG ARCHITECTS

COMPRISING

ADVICE TO THOSE WHO, WHILE YET AT SCHOOL, ARE DESTINED TO
THE PROFESSION
TO SUCH AS, HAVING PASSED THEIR PUPILAGE, ARE ABOUT TO TRAVEL
AND TO THOSE WHO, HAVING COMPLETED THEIR EDUCATION,
ARE ABOUT TO PRACTISE

TOGETHER WITH

A MODEL SPECIFICATION

INVOLVING A GREAT VARIETY OF INSTRUCTIVE AND SUGGESTIVE
MATTER

By GEORGE WIGHTWICK, Architect
AUTHOR OF "THE PALACE OF ARCHITECTURE," ETC., ETC.

Fifth Edition
REVISED AND CONSIDERABLY ENLARGED
COMPRISING TREATISES ON THE PRINCIPLES OF CONSTRUCTION
AND DESIGN

By G. HUSKISSON GUILLAUME, Architect
AUTHOR OF "THE THEORY OF ART," "DESIGN, CONSTRUCTIVE AND ÆSTHETIC," ETC., ETC.

LONDON
CROSBY LOCKWOOD AND CO.
7, STATIONERS' HALL COURT, LUDGATE HILL
1887

AUTHOR'S PREFACE.

THE Author has been frequently asked, "What should be the later school education of a youth intended for the architectural profession?" In answer to this he would refer to the earlier pages of his work as in themselves alone of much importance, in aiding the *parents* of such a youth to form some judgment as to his fitness, and the proper studies for his last year at school; while to *him* the entire book is useful in affording a general idea of those numerous acquirements, and of that untiring study, which may happily disabuse his mind of the notion that a mere aptitude for drawing is all that is required of him. This book will enable the parent to test the son's earnestness in reference to the scientific and arithmetical necessities of the practising architect, too frequently regarded with distasteful impatience by pupils when they come to the study of construction and the drudgery of cross-multiplication, the composition of a specification, and the computation of an estimate.

In fine, to the young man who *is* an architect, the practical portions of this work must needs be of much value, in at least directing his attention, and methodizing the process of his labours; while the general perusal of it will either confirm a youth in his determination for the profession, or serve to dissuade him from what might occasion only loss of time and useless cost to his parents.

PREFACE TO THE PRESENT EDITION.

ANOTHER edition of this enlarged and highly appreciated work having been called for, my aim has been to make it more worthy of the acceptance of the student and young architect, and of the favourable opinions that were expressed when the last edition was issued. In matters of fact few alterations have been rendered necessary; my attention therefore has been given to modify any crude expressions as far as the condensation necessary permitted. A note or two have been added on the change in architectural taste, and new matter substituted here and there in the Sanitary section. Parts IV., V., and VI. have, I need hardly remind the reader, been entirely added to the original work as left by Mr. Wightwick, and comprise those principles of science, construction, and architectural design with which the young architect should be familiar, and which it is hoped will lead him to pursue their study in other works of a more special kind. The new matter introduced in the original text is distinguished by being enclosed within brackets, thus [ ], or made as foot-notes. The model specification has been thoroughly revised, and considerable additions have been made to it. At the particular desire of many readers, and with the object of rendering the book more complete, I have appended a form of Conditions of Contract to the specification, which will be found to contain all necessary clauses. In its preparation I have consulted the forms approved by the master builders and council of the Royal Institute of British Architects and other authorities. Of course, special circumstances in every case will dictate to the architect modifications either in matter or phraseology.

G. H. G.

CONTENTS.

PART I.—SCHOOL STUDIES.

	PAGE
Natural Inclinations	1
Mathematical Studies	2
Drawing	4
Precision of Habits	6
Closing School Studies—Writing	7
Methods of Drawing	9
Method of Study	11
Hints on Form Delineation	12
Perspective	14
Perspective Projection	16
Hints on Perspective	21
Lights and Shadows	16
Cast Shadows	18
Isometric Projection	25
Technical Studies	28
Measurement of Builder's Work	29
French and Latin	30
Responsibility	31
Some Remarks on Education	32
Principles of Study	33
Hints to Pupils during Pupilage	37

PART II.—STUDIES ABROAD.

Travel	42
Italian Gothic	46
Time occupied in Travel	48
Hints to Architectural Tourists	49
Studies in France	52

PART III.—EARLY PRACTICE.

	PAGE
HINTS ON PLANNING	58
REFERENCE OF PLAN TO ELEVATION	60
ADVANTAGES OF PERSPECTIVE	61
ANGULAR PERSPECTIVE	63
SOLIDS AND VOIDS	66
JUNCTION OF HOUSE AND OFFICES	67
CHIMNEYS	68
CONNECTION IN COMPOSITION	69
POSITION OF WATER-CLOSETS	70
PLUMBERS' WORK	71
DRAINS	72
VIEWS OF EMPLOYERS	73
OCULAR IMPRESSIONS	74
VALUE OF CORNICES	76
WINDOWS	76
ARCHITECT'S DUTIES	79
CLERK OF WORKS	80
DETAILS	81
HINTS ON COMFORT AND CONVENIENCE—DOUBLE DOORS	83
POSITION OF FIREPLACE	84
WARMING AND VENTILATION	85
KITCHEN AND OFFICES	85
STABLING	88
LODGES	92
BUILDING COMMITTEES	93
ALTERATIONS AND ADDITIONS	94
QUANTITIES	95
DRAWING UP CONTRACTS	95
FORM OF AGREEMENT	96
GIVING CERTIFICATES	97
ARCHITECT'S POSITION	97

PART IV.—PRINCIPLES OF CONSTRUCTION.

SECTION I.—EQUILIBRIUM OF FORCES	99
MOMENTS	104
PRINCIPLE OF THE LEVER	106
CENTRE OF GRAVITY	108
SECTION II.—BALANCE AND STABILITY OF STRUCTURES	110
STABILITY OF WALLS, &c.	111
MODES OF FAILURE	113
THE LINE OF RESISTANCE	115
PRESSURE OF WIND	116

	PAGE
PRESSURE OF WATER	116
RETAINING WALLS	116
BRICK AND STONE ARCHES	118
ABUTMENTS	122
DOMES	123
STRENGTH AND STABILITY OF BEAMS	123
EFFECT OF POSITION	125
FRAMES AND TRUSSES	126
TRIANGULAR FRAMES	127
ROOF TRUSSES	129
ARCHED RIBS	130
STRENGTH OF TIMBER	133
JOINTS AND FASTENINGS	133
WEIGHT OF FRAMING	135
LOADS ON ROOFS	136
STRENGTH OF MATERIALS—TIMBER AND IRON BEAMS	136
DEFINITIONS	
TENSION AND COMPRESSION	138
LONG PILLARS	138
RESISTANCE TO CROSS STRAIN	140
STIFFNESS OF BEAMS	140
STRENGTH OF BEAMS	142
FLOORS AND ROOFS	142
RULES FOR SCANTLINGS	143
SECTION III.—IRON-CONSTRUCTION	144
CAST-IRON BEAMS	145
WROUGHT-IRON BEAMS	146
ROLLED-IRON BEAMS	146
STEEL	148

PART V.—SANITARY CONSTRUCTION.

SECTION I.—WARMING	151
FIREPLACES	151
CONSERVATION OF HEAT	156
HOLLOW WALLS	156
WINDOWS	157
DOORS	157
SECTION II.—VENTILATION	157
CIRCULATION OF AIR	159
SECTION III.—HOUSE DRAINAGE	161
SEWERAGE GASES	162
REMEDIES	163

	PAGE
Ventilating Traps, &c.	165
Closet Appliances	170
Sanitary Conditions of House Construction	171

PART VI.—DESIGN.

Section I.—Architectural Taste	173
Section II.—Principles of Design	179
Laws of Design	181
Mechanical Principles	181
Materials and their Functions	184
Iron and Timber	186
Wood-work	187
Plastic Substances	188
Cast Metal Work	189
Beauty of Form	190
Proportion—Unity—Variety—Contrast—Optical Refinement—Expression—Colour	196
Addenda on Construction	197

PART VII.—MODEL SPECIFICATION.

Index to Model Specification	203
Model Specification	213
Conditions of Contract	310
Miscellaneous Hints and Cautions	313
General Index	325

HINTS TO YOUNG ARCHITECTS.

PART I.

SCHOOL STUDIES.

It is my object to supply a course of hints which may prove serviceable, in the first instance, to the youth who is destined for the profession; and, in the second, to the young man who is about to enter upon the practice of it on his own account.

[NATURAL INCLINATIONS.—The direction of a young man's natural bent at so critical a period of his career as the last few years of his schooling, must in a great measure depend on a variety of considerations which a wise parent or guardian should well ponder. A young man's first choice is not always to be regarded as the right one. It may, as it often does, proceed from youthful aspirations which after-reflection shatters; from tastes which are based on mere sentiment; or from a notion prevalent among school-boys that some professions are exempt from application, hardship, or drudgery,—or that they open easy means of gaining a livelihood. The writer has frequently found the motive of choice to have been this hastily conceived notion, which subsequent knowledge or experience has dispelled, or which a due and careful consideration at the onset would have corrected.]

MATHEMATICAL STUDIES.—And, first, for the mere candidate who has yet to complete the last two years of his school studies. We presume that he has achieved a certain respectable quantum of classical attainment, with, at least, such a knowledge of the French language as it is now usual to afford in all well-ordered schools. Still cultivating these, it now becomes essential that peculiar care be given to the promotion of practical mathematics, geometrical drawing, and perspective. By the former we mean all that relates to the formation and measurement of superficial and solid figures, and those parts of arithmetic which have reference to square and cubical estimate and valuation; plane trigonometry, essential to the operations of the surveyor; and mechanics, necessary to compute the strength and strain of materials. By geometrical drawing is meant the free use of the compasses and steel pen, the drawing-board and T square, the camel-hair brush and Indian ink: this to be followed by an industrious application to linear perspective.

[The branches of mathematics more essentially necessary to the young architect, and which he should assiduously cultivate at school, are the following,—arithmetic as ordinarily taught, especially proportion, vulgar and decimal fractions, duodecimals, problems relating to square and cubical measurement, mensuration of lines, surfaces, and solids; geometry, or so much of Euclid as will give the student facility in constructing geometrical figures, erecting perpendiculars, setting off angles with or without the aid of instruments, besides a general knowledge of the more important theorems of Euclid; plane trigonometry, or the science of the relations subsisting between the sides and angles of triangles, a study of very wide application, applicable to all questions connected with constructive science as well as to surveying; and we may add an acquaintance with algebra, at least as far as the solution of simple equations. Among the branches of mixed mathematics the student should study mechanics; the principles known as the "Parallelogram of Forces," "Moments," and the "Lever," being particularly important and useful in the subsequent studies of an architect, as upon them the most elementary constructions in masonry, timber or ironwork depend. The elements of hydrostatics, hydraulics, and pneumatics should also be known, as many subsidiary questions involving such knowledge continually occur in the practice of an architect.*]

* To afford the student who is in quest of suitable works on the above subject a selection of some of the most desirable, we append the following list:—

Geometry.—"Geometry and Conic Sections," by Hann (Weale's Series). Todhunter's "Euclid." "Descriptive Geometry," by Heather: Lockwood & Co.

Trigonometry.—"Plane Trigonometry," by Hann (Weale's Series). "Trigonometry for Beginners," by Todhunter: Macmillan & Co. Griffin's "Algebra and Trigonometry:" Longmans.

Algebra.—Haddon's "Algebra" (Weale's Series). Todhunter's "Algebra for Beginners:" Macmillan. Griffin's "Algebra:" Longmans. "Integral Calculus," Rudiments of, by Cox: Lockwood & Co.

Mensuration.—"Mensuration," by T. Baker, published in Weale's Series: Lockwood & Co.; also "Land and Engineering Surveying," by same author: Lockwood & Co. Todhunter's "Mensuration for Beginners:" Macmillan & Co.

Mechanics.—"Rudimentary Mechanics," by Tomlinson: Lockwood & Co. For an elementary introduction, "Statics," by H. Goodwin, D.D., Dean of Ely (published in the Cambridge School Text Series), is recommended; or Todhunter's "Mechanics for Beginners." As advanced works on the subject, the student is recommended Twisden's "Practical Mechanics," Professor Rankine's "Mechanical Text Book;" "A Text Book of Applied Science," by G. H. Guillaume.

For students who require advanced Text Books on the above subjects, Gregory's "Mathematics for Practical Men" (Lockwood), edited by Mr. Law; or Professor Rankine's "Applied Mechanics" (Griffin & Co.), are highly valuable handbooks, as containing a digest of those parts of mixed mathematics essential to the practitioner. [In

Nothing is more common than for a young gentleman to enter an architect's office incapable of striking a circle without, at least, two ends; or of describing an octagon with any two sides alike; equally ignorant of cross multiplication—that leading essential of valuation practice, and bugbear of indolent reluctance; with no knowledge whatever of the use of the theodolite or spirit level; and having no idea that mechanics have any immediate reference to the permanent adaptation of stone and timber. A superficial reading of Euclid, and a course of algebra, may have gained a silver medal to be worn triumphantly on the last "breaking-up day;" but the peculiar application of the study to such matters as especially concern the architect will not have been thought of; and a thousand facilities, which might have been readily afforded before the day of apprenticeship, have been omitted to the great prejudice of subsequent pursuit. The self-flattering notion of manhood, natural to the emancipated youth, no longer a school-boy, is disgustingly corrected by the necessary incipient drudgery which makes him feel a child again—or leaves him the alternative of thinking himself too much a man for "task work" so elementary.

DRAWING.—His knowledge of drawing is illustrated by a series of rather free copies of very picturesque

In the subjects of Construction, Materials, &c., the student is referred to "Rudiments of Art of Building," by Dobson; the "Elementary Principles of Carpentry and Joinery," deduced from the works of Robinson and Tredgold; "Construction of Roofs," "Masonry and Stone-cutting," by E. Dobson; "Arches, Piers, and Buttresses," by W. Bland, in same series; "Limes, Cements, Mortars, &c.," by G. R. Burnell, C.E.; Humber's "Handy-Book for the Calculation of Strains in Girders," "Barlow on the Strength of Materials," edited by his Sons and W. Humber; "A Complete and Practical Treatise on Cast and Wrought Iron Bridge Construction," by W. Humber, Assoc. Inst. C.E., "The Application of Iron to the Construction of Bridges, Girders, Roofs, &c.," by Francis Campin, C.E., &c. The above works are published by Lockwood & Co., Stationers' Hall Court.

originals, in which there is but little of the formality of vertical or horizontal lines, and still less of lines perspectively convergent. Significations of trees, cottages, cows, and ploughmen home returning, all beautifully mounted, with gold lines ruled around, are exhibited to his future master, as proofs of certain removes from nature without any approach towards art. If he have any artistical feeling for *landscape*, the chances are he will not be *architecturally* inclined. If architecturally given, it is not unlikely that his "drawing-master" will have done his best to counteract the impulse. His geometrical drawing has been probably confined to a clumsy imitation of the figures of his Euclid, with letters that are *capital* only in a typographical sense. His writing:—ah, there indeed he flourishes! Words stretching out like race-horses, with long heads and tails raking into the lines above and below, so as to preserve a perplexing connection between whole sentences, past, present, and to come!

[We may add here, that a prevalent idea exists that "a taste for drawing" is the ultimatum or chief consideration in deciding on the architectural profession as the future calling of a youth. Greatly as it may afford facilities for the expression of ideas as the vehicle of design, it not unfrequently leads to a habit of cribbing prejudicial to that thorough independence of thought so necessary to the art of conceiving and working out the structural problems which engage the architect. A mere aptness for copying originals, drawing landscape and figures, while it may make an artist in the *Imitative* arts, must not be confounded with the creative function necessary to adapt various kinds of material for special purposes. It must not be forgotten that *design*, not drawing, is the main object, the *mind* not less than the hand being required.]

PRECISION OF HABITS.—Now, whatever may be necessary to other professions, or to any other branch of science or art, unquestionably there is no one which has more decidedly among its first principles the imperative law of *precision* than that of an architect, whether it regards the operations of the mind or the hand. The responsibilities which attach to him who may have to erect a large and important edifice, in which the economy of construction is to afford giant strength with graceful lightness, are such as should be considered from the very first moment of his architectural aspiration. *Precision*, then, in advancing, step by step, through all the gradations of initiatory study, demands the closest care. Architectural beauty is, in fact, the result of constructive perfection; and this can only be secured by laying down the first stone with a caution anticipating the pride that will attend the elevation of the crowning pinnacle. Each intermediate grade of operation will be also fulfilled with prospective and retrospective reference to all the others. The purpose and beauty of a building are indeed important; but the "very life of the building" is the foundation—*most* important, though afterwards to remain unseen. Many are the young architects who, on getting into practice, have had suddenly to make good with hasty, anxious, and health-destroying application, the omissions of their student years; and all, we maintain, from the early disregard of that precision of habit which, in the mind, means close and systematic study, and which indicates itself in the neatness and care of the hand.

But, while the young aspirant is not expected to have an intuitive sense of all this, neither is it to be supposed that he can, unaided, duly weigh its importance; and we therefore, through him, address those who have the care of his school studies after his profession may have been suggested. It will, at least, be unsafe, whether reasonable or not, to reckon upon any especial personal supervision at the hands of his professional master. His premium will be paid for the opportunities which he will have of learning—as his master learned before him; and in consideration of his ultimately securing to himself some of that remunerative practice which will be consequently forfeited by the senior professor. Moreover, it will assuredly be an all-sufficient answer for those who expect personal instruction, to say that all existing practitioners of any note have depended on their own employment of the mere opportunity for self-tuition.

It is, therefore, in regard to the duty which the student owes to himself (for he may, to a sufficient extent, do his master's business, and yet neglect his own) that we emphasise the necessity of a certain amount of school training preparatory to the deed of apprenticeship. During the term of his articles, we presume not to meddle with him further than to call on him to do his best, as he may rest well assured the more he serves his master the more he will serve himself. After that, we shall venture to take him up again; for circumstances may prevent his gathering from his more competent adviser those *extra-official* instructions which it will be our hope to afford him.

CLOSING SCHOOL STUDIES, WRITING.—To recur, then, to his closing school studies. We begin with the most simple.

The writing-master is first in request. The hand which usually wins the silver pen is about the worst that can be cultivated. We were ourselves nearly successful in the trial, and the mortification we experienced in having afterwards to curtail our capitals and control the comet-like eccentricities of our little *l*'s and *p*'s is not forgotten to this day. It was long before we could achieve the credit of being competent to write out the " fair copy " of a " specification ;" and many were the plans we rendered slovenly by a want of neatness and clearness in our figurings. The desire to give an official-like character to a "detailed estimate" was frustrated during almost the whole of our apprenticeship; and as to "printing" the titles of the fair drawings with the relative designation of the *plans, elevations*, and *sections*, there was a charity-school office-boy who ever maintained in this particular an envied ascendancy. Our good master used truly to say, that the writing and lettering would make respectable an indifferent drawing, and spoil a good one.*

The acquirement of a neat, close, and uniform character of writing, with practical ability in the several modes of lettering employed on drawings, is of immediate and of no unimportant use in an architect's office. The student, thus prepared, comes into the instant participation of advantages which otherwise he must wait for. The perfecting of his penmanship may be fully acquired at school, and he might there also so practise it as to make it still more essentially serviceable by copying out, for instance, a compendious architectural glossary, and giving to each leading word the particular letter which typographically expresses it. Thus, terms having exclusive application to classical architecture might be in the ROMAN character; those exclusively confined to gothic architecture in the 𝔒𝔏𝔇 𝔈𝔑𝔊𝔏𝔍𝔖𝔋; and all others of common meaning in *ITALICS:* the description being of course appended in the best kind of *Running Hand.* As a further

---

* We improved, however, marvellously, in our ordinary handwriting, and it was the mere penmanship of a letter of solicitation that a few years after obtained employment for us in the office of the late Sir John Soane.

exercise in the latter character, it would be well to copy out a dictionary of the technical terms of masonry, carpentry, &c.; the whole of which form no very lengthy operation, and would leave the practical illustrations of the office to be in their turn more interesting, because more readily comprehended.*

METHODS OF DRAWING.—The drawing-master comes next. What may be now his practice in schools we know not; but, in our own time, he did but very little in the promotion of artistical truth, and nothing at all in the way of practical utility. If the intended architect have—as indeed he should have—a feeling for the *pictorial* of art, it will not ultimately suffer under the requirement of its more anatomical and geometrical necessities. A decision of hand in outline delineation is the very first desideratum in an architectural draughtsman. This applies not less to curved than to

[* While we fully concur in the above suggestions, we may casually notice that the lettering now employed in most architects' offices has somewhat deteriorated in respect of the correctness and precision advocated by our author. In some offices, drawings are only written to in an ordinary hand; in others, a kind of fancy text is employed—a modified kind of German or Mediæval style. We append a few specimens:—

GATE-WAY : CATHEDRAL By

CHURCH : HOUSE :

NEW SCHOOLS ⚹ GROUND PLAN ⚹

S.W. View : MANCHESTER

Mouldings # DETAILS :

Longitudinal Section

In ground plans, sections, and general drawings, an intelligible block letter or plain italic lettering is recommended as more conducive to clearness and appearance than any ornamental or grotesque style.
[Wo

direct lines; nor can a better study be suggested than that of the human figure, beginning with the skeleton, combined with the representation of plain solids of regular form, without (in the first instance) any shading or attempt at *effect*.

When the hand has acquired some independent precision and firmness, the use of the mathematical instruments, drawing board, T square, and parallel ruler should be carefully attended to; and all this will be sufficiently induced in the study of perspective. The pupil will first draw the plan of his subject, then the elevation, and finally work out its perspective appearance under certain prescribed conditions.

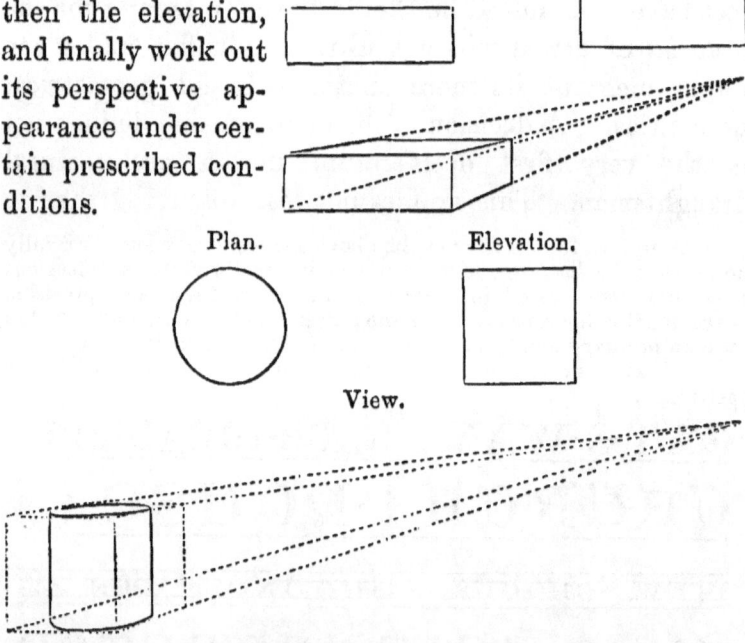

His previously required precision of hand is now regulated by the knowledge of rule; and both will work together to make a correct and ready sketcher. The study of the finished human figure (which should alternate with the more formal process), while in itself

We may remark here the strong feeling for or affectation of Mediæval character shown in the above specimens—a tendency which appears to approach the ridiculous in some instances, and is to be avoided.]

most valuable in respect to statuesque architecture will equally facilitate the drawing of mouldings and their enrichments; and we may here add, that a careful copying of the leading frieze ornaments, Greek and Roman and Gothic, will be infinitely more to the purpose than any attempts at that mediocrity in picture which never afterwards aspires to more than a place in my lady's album.

[METHOD OF STUDY.—Since our author's remarks were first published, the Governmental schools of art which have sprung into existence have materially improved the mode of instruction in drawing. Our author's drawing-master, no doubt a veritable recipient of art patronage in the past, has, happily for the advancement of art, given place to teaching, or rather a system that promotes the training of *art faculty* in contradistinction to the mere penmanship of the art. Instead of a series of pencil sketches or crayon drawings of landscape or the human figure being thrust upon the young beginner, he is taught by a graduated system of instruction to acquire a decision of hand in copying the simplest outlines and geometrical forms before he proceeds to the representation of the concrete picture in its complete detail and expression. He is thus, by a rational process, led to represent *form* and *solidity* under their varied phases of complexity and perspective appearance, and so to grasp intellectually and ocularly the changes which objects undergo in Nature by their relative position, and the effects of light and shadow. By these means the student really learns not only how to copy Nature perfectly, but how to present her to the best effect. We would suggest here to the student who is pursuing this branch of his education to trust more to his eye and mind than to his imagination. He will thus learn never to draw a line unless he understands its effect, nor to make a touch where it is not needed. Copies from drawings are to be steadfastly withheld, until facility in expressing the simplest outlines and solid forms is acquired. We would earnestly recommend the following method of study:—

1. Take the easiest and least tedious forms first, *e.g.* lines vertically, obliquely, and horizontally drawn; triangles, squares, parallelograms, and polygons; curvilinear figures, circles, ovals, ellipses. These representations should be made first as planes.

2. The effects of position in relation to the eye should next be studied as a preliminary and tentative to scientific perspective.

3. The solids—as cubes, pyramids, prisms of various bases, cylinders, cones, and polyhedrons, may follow. These should first be drawn geometrically with regard to the eye, and afterwards in different oblique positions and levels, perspectively.

4. The effects of light should next be carefully studied, the cast shadows being drawn in outline, and then the varied gradations of shading and reflected light upon solids, as octahedrons, cylinders, spheres, cones, &c.

5. Plain shading in pencil, Indian ink, or sepia may be succeeded by flat washes of colour, artistic accessories and finishes. In a word, begin with the *simple* and proceed to the *complex*, the *abstract* and then the *concrete*. Model drawing or construction, or real object delineation, should always precede or be combined with the generalised effects of Nature, or the complete picture. Sketching directly from Nature, however crude the first attempts may be, is recommended the student in preference to any course of flat copies.

HINTS ON FORM DELINEATION.—We offer a few hints on the delineation of those forms which puzzle the beginner in drawing, and which are of perpetual recurrence in architectural design. To assist the eye and hand, the student should endeavour to grasp the proposed form constructively, *i.e.* to reduce it if possible to a simpler outline; thus a circle generally perplexes the novice, and few even expert draughtsmen can draw a perfect circle merely by the hand without aid. The eye requires some means of judging distances and boundaries, and hence it is easier to inscribe a circle within a square than without such help.

Let it be required to draw a circle of a certain diameter. Denote the centre by a point. Through this point draw two diameters crossing each other at right angles, and mark by the eye as nearly as possible the required radius on each from the centre; if further aid be wanted, the four equal right angles may be

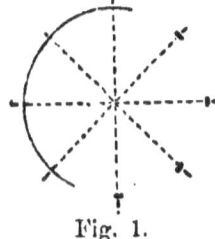

Fig. 1.

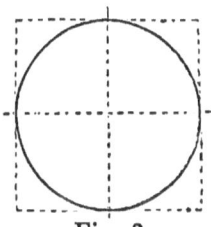

Fig. 2.

bisected by two other lines, and radii marked as before (see fig. 1). The hand can now trace the curve. A circumscribing square is another good plan, as in fig. 2. When a circle has to touch a certain boundary, as in the circle of a traceried window, unite the points of contact by diameters, as in fig. 3. To draw a polygon by the eye: If it contain a regular number of sides, as an

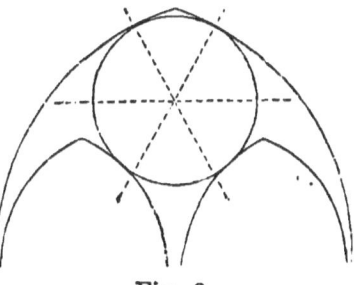

Fig. 3.

octagon, draw four intersecting diameters as for a circle, and mark equidistant points from centre.

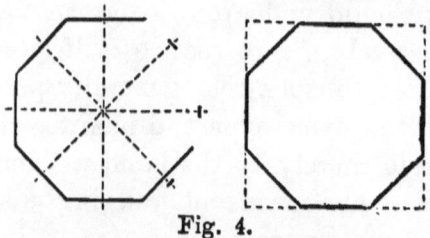

Fig. 4.

Then draw at right angles through each point or extremity the respective side; another plan is to form a circumscribing square, cutting off equal distances by the eye (fig. 4).

CURVES.—Segments of circles and curvilinear forms frequently perplex. In these cases fix by the eye the

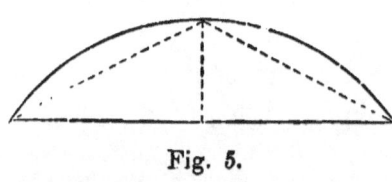

Fig. 5.

limits of space or chord line and the rise (fig. 5), or circumscribe the boundaries by rectangular lines. (For further remarks on curved lines, see section on Design.)

EYE PERSPECTIVE.—To draw forms perspectively by the eye. Consider the level and position of the eye in respect to it. Draw a horizontal line to denote the former, and a point in it for the latter; from the boundaries of the form draw convergent lines to the point of sight. Thus, to put a circle in perspective, either horizontally or vertically, draw the converging lines and set off the width as nearly as can be judged by the eye, thus forming a perspective square. Draw the curve touching the centres of the respective sides, as in fig. 6.

An octagon is drawn also by cutting off the angles of a perspective square (fig. 7). Cylinders and other solids are drawn by first forming the circumscribent parallelopidedon by combining six perspective squares, and then forming the required solid by the rules given above. Remember that circles become ellipses, and squares oblongs in perspective.

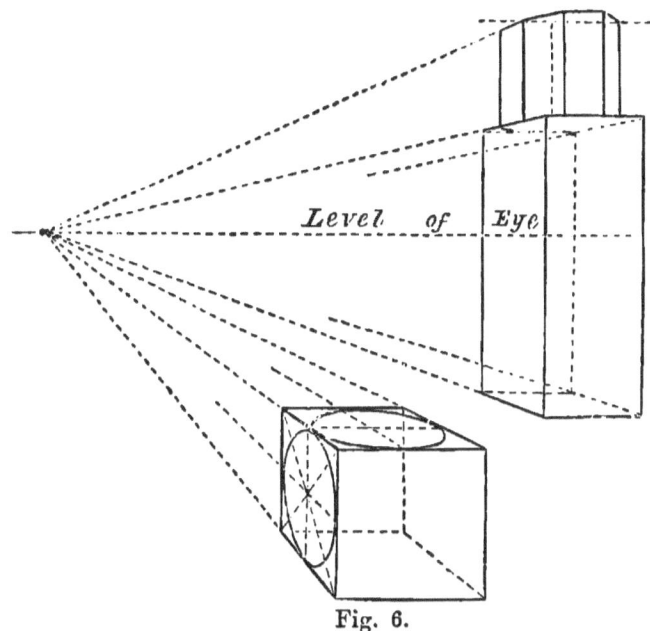

Fig. 6.

These hints are simply in reference to eye drawing or sketching, and assume no instrumental aid; but of

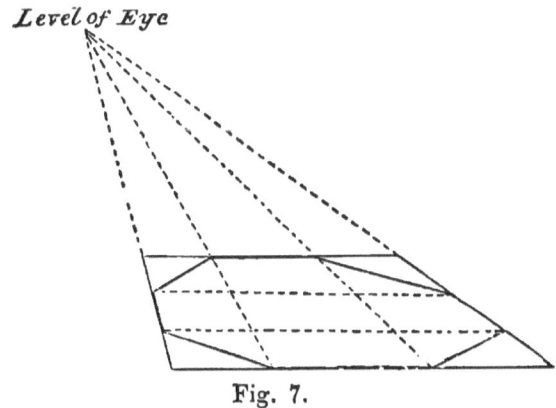

Fig. 7.

course the student destined for the profession should acquire from suitable treatises the systematic principles of geometrical drawing, Projection and Perspective.

PERSPECTIVE PROJECTION.—Perspective is a word which excites a dread or despair in some minds, but without just ground. It is nothing more than a knowledge of form in its relation to the sense of sight, or of the visible effects of lines and planes in different positions, and the means of representing them on flat surfaces, as paper. Many young students are inclined to halt at what they imagine requires great trouble and mathematical skill. The effects of Perspective are either linear or aërial, the first depending upon *position* and distance, the latter upon distance or atmospheric agencies. It is necessary the student should remember that all receding surfaces are modified both in shape and tint, or diminish by the laws of perspective to certain points in the horizon opposite the eye called the points of sight and distance, and also that as they recede they fall into a half-shade.

LIGHTS AND SHADOWS.—The parts of surfaces nearest the illuminating power receive the light, and thence graduate through a series of half-tints to the opposite or farthest side which falls into shadow. It is to be noticed, however, that these extreme shadows are modified considerably by reflected light from objects near, and therefore it must be observed that the *extreme* outlines of an object seldom receive the brightest light or the deepest shade (see fig. 8). All unpolished surfaces as they recede become darker, or fall into "half-tint," thus retiring surfaces are indicated pictorially by shadow. To illustrate these principles: Horizontal planes recede more from the eye than many vertical surfaces, though they generally receive greater illuminating effect, and must therefore be modified accordingly. But when the upper surface of

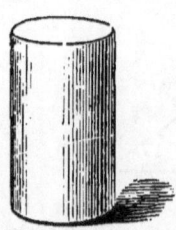

Fig. 8.

an object is nearly level with the eye, it will recede more rapidly than the front retires from the light, and the latter should be made lightest (fig. 9).

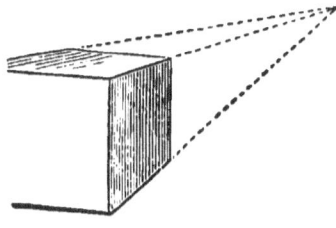

Fig. 9.

Receding vertical faces fall into half-tint or shade still more by receding both from the light and from the spectator (fig. 10). Surfaces *inclined to the light*, as the side of a roof or front of a building, receive more light directly than any other sides, and the contrast between the lighted and shaded sides, as when an angle is presented to the spectator, produces a stronger shade nearest the angle between them, as in (fig. 11). This is chiefly due to the effect of contrasted light and shade, and

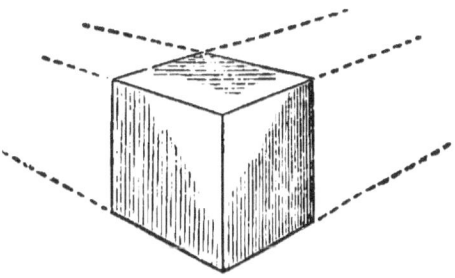

Fig. 10.

is useful in producing architectural effect. Surfaces which are perpendicular, as walls, and recede, will be lightest at the upper corner nearest the spectator, graduating thence into half-tint, and if in shadow, the nearest edge will be the darkest.

Cylindrical surfaces, as columns, receive reflected light on the shaded side and close to its boundary, and the highest light at a corresponding position on the opposite side,

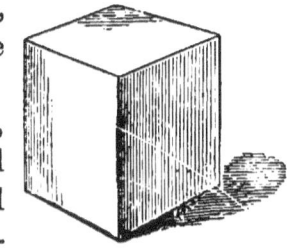

Fig. 11.

the deepest shade being attained between; graduating in delicate tints to the extremes (see fig. 8). Concave surfaces, as niches, domical ceilings, &c., are to be treated similarly, except that the light and shaded sides are reversed, the deepest shade being nearest the light, the half-tints and reflected lights being graduated between; the extreme edges receiving either a delicate half-tint or reflection. The reflections on a concave surface are stronger than on convex surfaces.

CAST SHADOWS are shadows thrown by objects intercepting the light in the direction opposite to that of the light. Cast shadows are modified variously by reflection, atmospheric refraction, source of light, opacity or transparency of the objects casting them, surface upon which they fall, &c. When the light proceeds from a lamp or flame of a candle, the shadow is strong and defined; but when it is from the sun the shadow is modified by reflected light. Again, the shadow is stronger in proportion to the nearness of the object to the source of light, and is more defined upon that portion of the ground or surface upon which it falls which is nearest to the object. Shadows are modified in intensity by their contrast with the lighted surface or ground. They are deepest and most defined when in contact with the brightest light, or that part of surface which receives the greatest light.

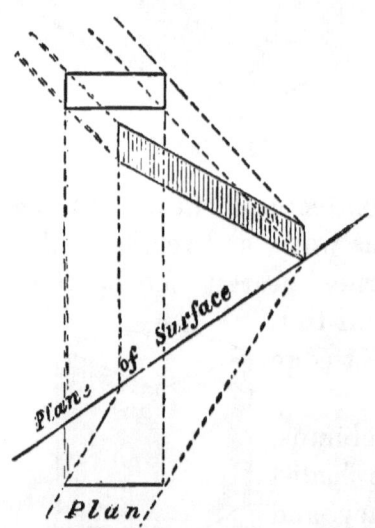

Fig. 12.

CAST SHADOWS PARTAKE OF THE FORM OF THE SURFACE UPON WHICH THEY ARE THROWN.—Flat surfaces receive the form of the object, modified only according to the angle which the rays of light make with the plane. Thus, when the plane is perpendicular to the rays of light, and the object interposed also perpendicular or parallel to the plane, the shadow resembles the shape of the object, and when cast by the sun, also in size. When the plane is inclined, or the object is obliquely situated in regard to the plan, the shadow is increased in proportion to such obliquity (see fig. 12).

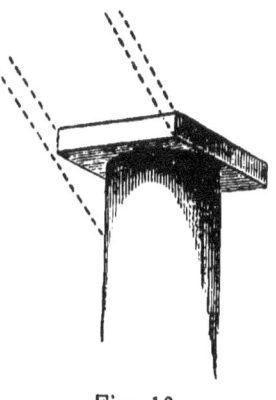

Fig. 13.

Upon convex surfaces, the parts which are not directly perpendicular to the rays or the receding portions, produce an exaggerated shadow, as the shadow cast by the abacus of a column upon the circular shaft which assumes a concave or elliptical outline, unless the rays of light and the eye of the spectator are coincident (figs. 13, 14). The curvature of the shadow will be greater or less as the angle formed by the rays of light are more or less oblique.

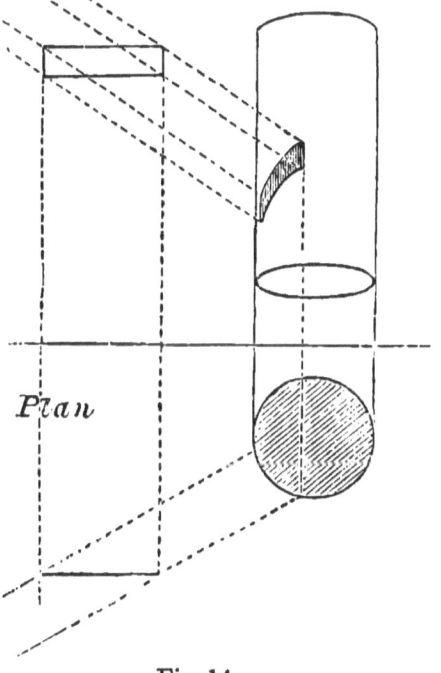

Fig. 14.

In concave surfaces these effects are reversed, the

deepest part of the shadow being cast upon the extreme depth of the concavity.

Upon irregular surfaces made up of projecting and receding planes, convexities and hollows, the form of the shadow partakes of the varied surfaces (see fig. 15).

It must be observed that the value of cast shadows is to give *relief* and *express the surfaces* upon which they are thrown. As a rule, it is preferable that they should follow the projections and irregularities of the surfaces

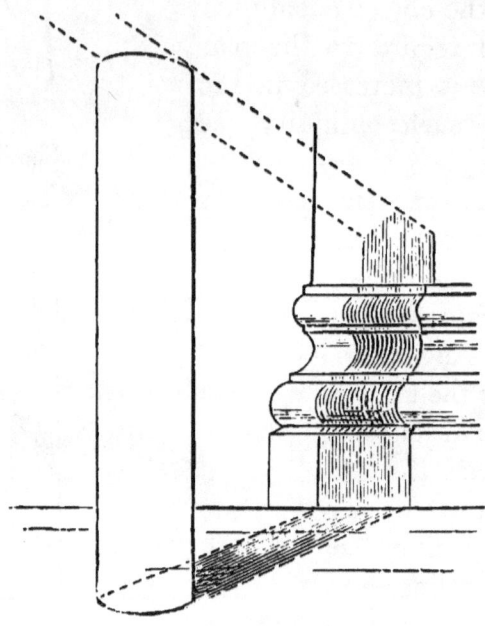

Fig. 15.

simply, and thus become expressive of ground or other surroundings of the objects. In cast shadows of towers, pinnacles, chimney-stacks, &c., upon the inclined surfaces of roofs, this rule becomes of much value, as affording evidence of projection and depth; though of course regard should also be had to the shape of the objects themselves. Sometimes, in the cases of curved or moulded surfaces, the only evidence or description of such parts can be given by bearing in mind the above rule, and hence for architectural reasons it is to be preferred. This kind of projection and relief becomes, when *scientifically* employed, a valuable accessory to the architect, though it is too frequently, from ignorance or carelessness, or what is more censurable, a love of deception, misemployed, or *worse* than useless.

HINTS ON PERSPECTIVE.—We do not propose here to write a treatise on the art of Perspective, but simply to aid the student in his comprehension of it, and to lay down a few rules.

Linear Perspective has for its object the determination of various points and lines of given objects upon a plane surface, and is purely a geometrical projection. It is really the projection of the *section* of a pyramid whose vertex and base are given—the first being the eye of the spectator, the latter the boundaries and surfaces of the objects to be placed in perspective; the *cutting plane* is the picture.

If in fig. 16, A, B, C, D, E, be the plan or horizontal plane of projection of a pyramid of which A'', b, c, d, e, is the vertical projection; and if T'' t T' be the traces of the vertical plane of the picture in these respective planes; let it be required to find the position on the picture of any point as a', or a'', these being the points where the vertex A of pyramid cuts the traces of the picture plane T'' t T'.

Fix on the picture two lines as axes X X and Y Y to which to refer all other points, these axes being fixed also on the planes of projection. Let them be chosen to pass through the eye, or as representing two planes drawn through the eye, one horizontal and the other vertical, and both perpendicular to the picture as O X, O Y. These lines will cut the picture planes in x and y respectively. Then since these axes are determined both on the picture and on the planes of projection, it is easy to determine the position of *a'* both *horizontally*, namely, *y a*, which will give the vertical for the vertex of

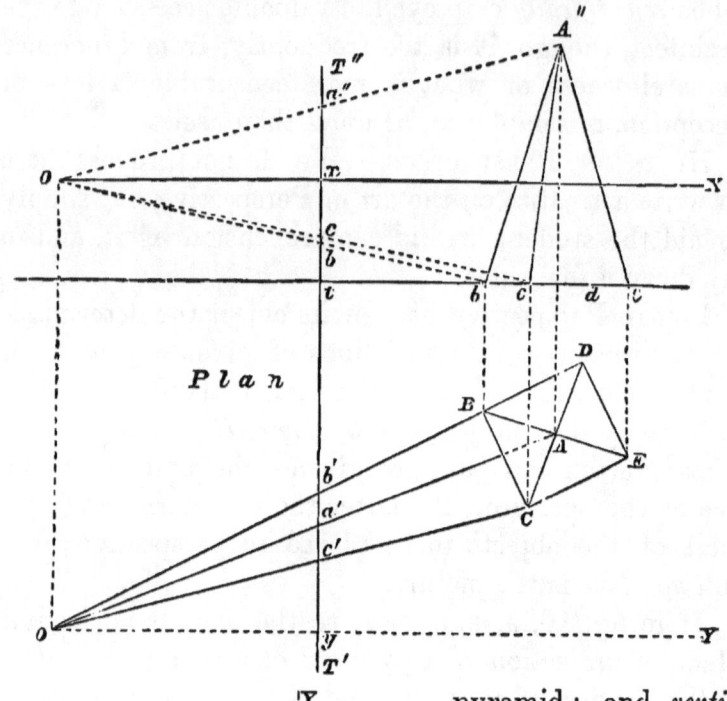

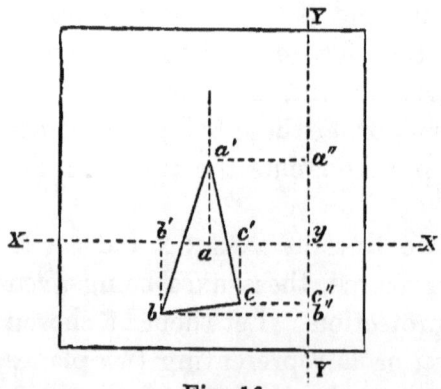

Fig. 16.

pyramid; and *vertically* also, since *x a"* gives the distance from the axis x x on the picture. Thus the distances from the respective axes x x and y y are determined and the point *a"* fixed. Any other point may be determined in the same manner, and the pyramid completed.

Theory of Lines and Planes.—The perspective representation of a straight line is a straight line in every position, and it is sufficient to determine the perspective positions of two points in it and join them.

Planes or lines parallel to the plane of picture have no vanishing points, or are assumed to have none.

Planes or lines inclined to the picture plane vanish in that point where a parallel line, drawn from the eye, would meet the same plane.

All planes and lines parallel to each other but not to the picture plane have a common vanishing point.

The perspective representation of all vertical lines will be parallel and vertical.

The centre of the picture is the vanishing point of all lines perpendicular to the picture plane, and is always where a perpendicular line from the plane would project and touch the eye, or where a line drawn from the eye perpendicular to picture plane meets it, and is called the "point of sight."

The distance of the point of sight or station point *from the picture* when placed on either side upon the horizontal line, is the vanishing point of all lines inclined to the picture at an angle of 45°.

The line drawn from the eye of a spectator to the plane of projection, gives a vanishing point, and is always parallel to the plane of the object for which it is the vanishing point. All parallel sections produce similar figures; thus all circles situated in vertical or horizontal planes (not parallel to the picture plane) are ellipses; all squares so situated become oblongs, diminishing in width in proportion to the acuteness of the angle they make with the line of sight.

Spheres.—The perspective representation of a sphere is an ellipse, except when the line of sight coincides with the centre of the sphere, when it is a circle. This is evident from considering the definition of perspective before laid down, namely, that the cone of rays proceeding from the outline of any object as a globe if cut by a plane (as the picture plane) obliquely, the section produced will be an ellipse. A globe thus presents its longest diameter in a direct line to the eye if the rays are cut obliquely.

Columns standing in a line parallel to the plane of picture, will, according to the rules of perspective, be of equal heights, but their diameters will appear greater on the said plane as they recede from the eye to the right or left.

The two latter theorems appear paradoxical, though by the arbitrary definition of perspective, which supposes lines parallel to the plane of delineation not to vanish as they recede from the station or point of sight however distant, are quite correct. Few artists would, moreover, think of making such a row of columns present larger diameters as they receded; nor would they tolerate making globular forms spheroidal. A little consideration, however, will convince the student that such an assumption is purely conventional, and made for simplicity's sake. It is evident distances, whether vertical or horizontal, must equally diminish as they recede from the eye of the spectator, whether they be geometrically presented or not. Perspective, as we have said, assumes planes and lines parallel to the picture to be unaffected or not to vanish, and consequently cannot take cognizance of exceptions; but, correctly and theoretically, the assumption is erroneous, as all distances, and of course diameters of cylinders and globes, must unexceptionally obey the law of diminution which perspective teaches.* This is certainly

---

* If we suppose the cutting plane or picture plane to be equidistant from the eye of the spectator everywhere, or the concave surface of a hollow sphere instead of a plane, all distances would be correctly represented on it, as the visual rays would always be cut at right-angles instead of obliquely.

an inconsistency in the theory of perspective as usually practised, though the practical inconveniences resulting may be small. Correctly also, by the same law, vertical lines, as the lines of towers and lofty buildings, must vanish upwards to a point perpendicularly above the point of sight, and long façades should for the like reason diminish laterally, even when they are parallel to the plane of picture. The latter theory implies a flat catenary or hyperbola, or certain points in the horizontal lines in the vertical plane of the eye where the convergence begins. Such considerations, however, are not practically of much value.

The problem presented by perspective reduces itself to this; namely, the construction or determination on the picture plane of the point in which the visual ray drawn from the eye to a particular point of the object meets it. It is for the draughtsman to use in each particular instance the least troublesome means of resolving it, which may be considered as follows:—

Conceive two different planes drawn through the proposed point to be put in perspective and the eye, the visual ray being their intersection; if these planes or their lines of intersection with the picture plane be drawn, then the point of intersection of the visual ray with the plane of picture will be obtained also. The draughtsman can choose those planes which can most readily be drawn through the eye and the point to be placed in perspective.

ISOMETRIC PROJECTION.—One of the most useful and elegant systems of projection applicable in an especial manner to architectural constructions and groups, may be noticed here, though seldom used by the architect to the extent its value merits. The principle of this system, called isometric projection, invented by the late Professor Farish, of Cambridge, consists in selecting for the plane of projection a plane equally inclined to three principal axes at right angles to each other. These are called the *isometric axes*, and all lines and planes parallel to them are representable to the same scale. The lines are called *isometric lines*, and the planes the *isometric planes*.

The isometric projection of a cube will convey the principles of this system.

Let A B, A C, A D be the isometric axes, and A the "regulating point," or origin of the axes; then the plane of projection makes equal angles with them, as also with the other edges parallel to these, and is perpendicular to A F, the diagonal of cube; the projection of this diagonal being the point A, the projections of the edges of the cube will be equal lines. Also the edges or axes A B, A C, A D being equally inclined, the angles between them are equal, being each equal to 120°.

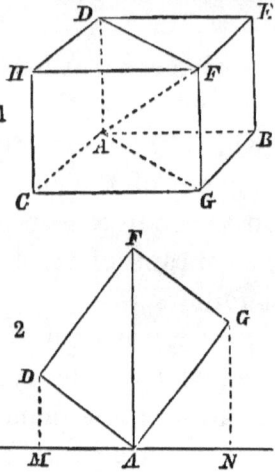

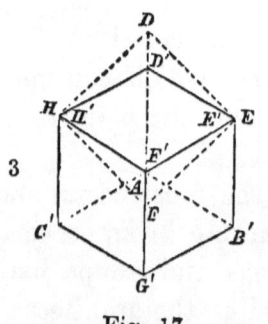

Fig. 17.

Moreover the projections of all the diagonals from the points A and F are coincident with the projections of the cubical edges; and the points B, C, D being equally distant from the plane, the diagonals joining these points are parallel to the plane of projection, and their projections are consequently equal to the diagonals. It is evident also that the angles which the projected edges make with the diagonals of each of the three projected faces of the cube are each equal to 30°.

To construct the projection of a cube isometrically. Let F E D H be the face of the cube and H E its diagonal. From E and H draw E C' and H B', making angles of 30° with H E. These lines will meet in F'. Then complete the upper face of the cube F E' D' H'. Produce a diagonal D' F' to G', making it equal to one of the axes F' E', F' D' or F' H', complete the adjacent sides of the parallelograms, and the projected sides of the cube are completed. The projections of the farther edges of the cube are coincident with the diagonals already drawn.

Along the isometric axes may be set off by any given scale the dimensions to be represented in the directions of those axes. Thus distances can be set off along all isometric lines as required.

If the length of an edge of the cube be 1, the diagonal of a face is represented by $\sqrt{2}$ and the diagonal of the cube by $\sqrt{3}$. Then since A F : F D : : A D : A M (diagram 2, fig. 17); A M the isometric projection of an edge of the cube $= \dfrac{1 \times \sqrt{2}}{\sqrt{3}} = \tfrac{1}{3} \sqrt{6}$
$= \cdot 81649.$

The student is recommended to practise himself in this useful kind of projection by putting the blocks of any buildings into isometric perspective; for designs of unions, hospitals, prisons, colleges, and other assemblages of detached buildings, it offers many advantages of delineation, enabling a better idea to be formed of enclosed spaces and areas, and answering the double purpose of a block and roof plan. It is sometimes called " bird's eye " perspective.*]

---

* The student should study Jopling's "Isometric Perspective." Descriptive geometry should also be carefully studied, as aiding him in adopting the easiest and most practical methods of projecting solids upon flat surfaces. The chapters on "Stereography," in Newland's "Carpenters' and Joiners' Assistant," afford also elementary knowledge.

A certain advance being thus made in linear drawings (which, by the way, having been first carefully pencilled, should then be more carefully perpetuated with the steel or reed pen in Indian ink), the camel-hair brush will be introduced. To produce a smooth and uniform tint, accurately terminating with the boundary lines, is the first step: to practise the "softening off" necessary to give effect to a circular pillar or a sphere, the next. The distinctions between positive and reflected light, shade and shadow, will succeed; and then will follow the leading rules by which shadows are accurately projected. After this, it will be time to indulge in the charms of colour; and when the student has mastered the means of dealing with superficies and solids, his drawing-master may then do his best to make an artist of him.

[The late Mr. Peter Nicholson's, and the many other treatises on architectural drawing, projection and the mechanical principles of architecture, have placed within the reach of school tutors means of instructing pupils destined for this profession, which, if neglected for the dead languages, betrays on the part of the tutors a want of discernment, and becomes a serious hindrance to their pupils.]

TECHNICAL AND MECHANICAL STUDIES. As a mere act of mental discipline, a course of the classics will be of equal value with mechanics; but the practical value of the latter will be at once apparent when it is considered that the means of constructing every building involve the use of all the mechanical powers, and that a knowledge of the properties of the simplest of these—the lever—is of the utmost importance as affecting the laws of the resistance of timber, the composition and resolution of forces. The construction of roofs, overhanging partition framings, and trussed beams, becomes a matter of prominent interest when the mechanical principles on which it depends are understood; and all we contemplate, in speaking of the young architect in embryo, is the guidance of his mind into that train of practical thought which the obvious *utility* of his early studies will promote. There is nothing to prevent a school-boy from mastering such brief elementary articles as those of Mr. Gwilt on " the Equilibrium of Arches " and the necessary magnitude of their piers; or such as Mr. Bartholomew's Chapters on Gravity, "the source of all principle and defect in architectural construction;" nor can we resist an allusion to the articles on masonry and carpentry published in a separate form from the " Encyclopædia Britannica."*

MEASUREMENT OF BUILDERS' WORK.—Allusion has been made to the arithmetical and mathematical studies of the intended architect. It is most due to the value of his time when he first enters the office of his professional master, that he should be already fully practised in all that relates to square and cube measure; quick and sure in the working of cross multiplication, and in carrying out the sums, or what is termed the "moneying" of quantities at different prices. It will not be his business, until he is apprenticed, to learn the different ways in which the differing branches of artificers' work are measured; or their value *per* yard, *per* foot, *per* rod, or *per* perch, superficial or cube; but he ought to have at his finger ends the mere calculative process, so that his attention may be given, from the first day of his entering an office, to acquiring a knowledge of the nature and value of labour and materials, and of the varied way in which the mason, bricklayer, plasterer, carpenter and joiner, plumber and painter,

* The student is referred to Part IV. of this work for the elementary principles of the above subjects; and the rudimentary treatises on " Carpentry, Masonry," &c. (Lockwood and Co.)

compute their perfected operations. All the ordinary rules of arithmetic are of course of absolute necessity. We only speak of those in reference to which the *utmost practical readiness* is immediately required. He who can square the greatest number of dimensions in the least possible time, in like manner multiply quantities by prices, and "add up" a foot long of closely written pounds, shillings, and pence, has received, as it were, an impetus which will carry him upwards on the ascent of early practice, and vastly alleviate the tedium of this most imperative drudgery.

His mathematical studies should in the same degree facilitate the construction, with his compasses and steel pen, of circles, squares, triangles, pentagons, octagons, and so on to polygons, divided and subdivided,—of parallels that have no chance of meeting, and truncated isosceles triangles that have their apex in some unattainable point beyond the other end of the school-room. Trigonometrical practice, too, might be carried on at least so far as to produce a map of the play-ground, including "the duck-pond and three elm-trees beyond," with the respective levels of the several angle points. He should be enabled to raise a perpendicular line without his drawing board and T square, and to draw a raking line at any required angle. Of the many things which are generally only touched upon at schools, such leading ones as we have mentioned should be fairly grasped. A three years' term would be thereby rendered equal to the usual five; and, at the end of the five, a salary might be commanded, where a mere interval of anxious incompetency is the common instigation to forfeit *more* time, in making good the loss of time past.

FRENCH AND LATIN.—We have spoken of some knowledge of French as usually afforded at schools.

The more of it that can be there attained, without injury to the immediate essentials before enumerated, the better; because, in conjunction with Latin, it promotes those facilities of travel of which we sincerely hope our intended young architect may be enabled to avail himself. At all events, then, what *can* be learned at school should not afterwards be forgotten; and this prompts us to make the only reference in which we shall indulge as to the duty of the student during his apprenticeship. After Latin and French, *Italian* is easily acquired. The study of this simple and beautiful language (speaking of it as a medium of communication, and plain literary instruction) will prove a mere pastime. The writer of these hints (though naturally slow at languages) learned sufficiently, by two months' *exclusive* application, to read simple prose readily, and to make his way through Italy tolerably well.

RESPONSIBILITY.—Lastly, we would impress upon the young aspirant to architectural honours, our repetition of the *responsibilities* which will attach to him from the first hour of his unaided practice. It may be some time before he will be enabled to purchase assistance; and during that state of individual probation, he will have—if he have employment—duties relatively more arduous and more harassing than when commissions shall thereafter pour upon him to the hoped-for advancement of his fame and fortune. He must be for a time " grand master," assistant surveyor, and drudgery clerk, of his own establishment: at once designer artistical, constructor practical, copying draughtsman, measurer, valuer, and more—with which we would not frighten him. He must cultivate resolution on the ground of knowledge, endurance on that of patience, and modesty in the full assurance, that, when he shall have practised to the last day of his occupation, he will have learned the more to know how much he has yet to learn. His profession is a noble one, based on palpable science, and beautified by the poetry of art. It is most gratifying in respect to the society to which it may lead, and the rank it may confer. It is more especially so in regard to the pride which an architect cannot but feel in contemplating the material and enduring majesty of the structures he may have to raise. Paintings must be sought in the gallery; statues may indeed preside in the open square; but it is architecture only which towers into the sky—alike commanding, far or near; and combining the graces of form, proportion, and decoration, with picturesque charm and massive grandeur.

We now take leave of our school student, to meet him again some five years hence.

[SOME REMARKS ON EDUCATION.

We cannot pass from this initial stage of the student's career without offering a few observations. If our school training in some respects has undergone a change for the better since our author's time, it has also a great deal more to accomplish, in the way of rejecting an unnatural system of education, and helping the spontaneous development of the mental faculties. From some experience in the teaching of young men, and the observation and reflection which have been the result, as well as the study we have given to this subject, we offer these remarks. An able and philosophic writer observes the maxim of the school has been, "believe, and ask no questions." Spontaneous activity of mind has been too much checked in the past, and parents and teachers are beginning to understand that the tastes and inclinations of pupils must, to a larger extent than supposed, preside over and determine the future, as well as the studies and kind of training, of each mind. In its restrictions and dogmatic notions, the old *régime* of education and society have been exactly counterparts. It is true, also, our modern ideas of culture have changed with our religious and political ideas. The grammar-school routine of half a century back placed implicit faith in authority and classical oracles; and a fictitious reverence for the gods of fable and history asserted a supremacy over all the more natural instincts and faculties of the scholar. The decline of such tutorial authority is the natural result of that slow rebellion of intellect which has lately given rise, in its turn, to new systems, and the accumulation of methods. Such a diversity of methods and systems cannot fail by a general law to approximate towards a truer course. Let us briefly revert to the errors of our forefathers in this important subject, and lay down the principles of a more rational method of study which the architectural student especially should be led to follow.

PRINCIPLES OF STUDY.—1. *The reciprocal action of mind and body.*—It has been truly observed by our greatest physiologists and psychologists* that body and mind must be both cared for and sustained to their mutual help. The old school *régime* notably was one-sided in this respect, giving an undue prominence to book-learning without perceiving the necessary co-operation of the physical and mental powers; hence we find the reaction of one leading to extravagance in the opposite direction. The old method of "forcing" and rote-learning, the mechanical way of teaching languages by mere rule and repetition; the absence of technical instruction, and the substi-

---

* See Dr. Spurzheim on "Education;" Herbert Spencer's; and Professor Bain's work on "Body and Mind," &c.

tution of mere formulæ and symbols, are now happily beginning to be superseded. Sign-knowledge instead of facts, and correct literalism instead of understanding, were evils which led to a neglect of essentials both in literature and the arts. Montaigne has said, *Sçavoir par cœur n'est pas sçavoir.* The mind absorbed by the mere symbols of knowledge cannot attend to the *things* signified. Herbert Spencer aptly says, "Between a mind of rules and a mind of principles there exists a difference as that between a confused heap of materials and the same materials organised into a complete and compact whole." We cannot impress this too forcibly upon the student of architecture, whose school-taught rules avail him little when confronting the actual facts of practice, when the special requirements of each case demand a wider and more generalised knowledge. A mind limited to rules is at sea in emergencies.

One phase of the question especially concerns young architects. It is that the mere symbolism of art-knowledge—"styles" or modes of art expression—has been the wreck of architecture during the last four centuries. Instead of studying principles, architects have been studying rules and forms *ad nauseam,* the consequence being they have been obliged to copy and repeat the past. In one word, the mind requires experimental acquaintance, or the senses to go hand in hand with the mind. This leads us to our second principle.

2. *Particulars should precede generalisation.*

The substitution of *principles* for rules is now admitted to be the right course.

The inquiry and independent thought of the pupil is awakened by the calling up of experiences and facts. The mere verbal envelope of ideas is of little use unless accompanied by previous experiences learnt through the senses. The scientific formula must follow the attainment of simple facts; in other words, the particulars should be learnt before the general truth. Science is, in fact, a collection of rules, the result of induction and comparison of facts. Neither nations nor individuals arrive at science first. Hence the absurdity of making the grammar and syntax of language precede the acquirement of ideas and words. The young pupil should thus get an accurate knowledge of the visible and tangible properties of form, quantity, weight, &c., before his inferences can become just or methodical. Especially is this the case in the learning of mathematics, geometry, mechanics, and the like. With the artist no less than the physicist, the old *à priori* method of enquiry should give place to inductive notions of form, number, and a correct diagnosis. The relationships of scientific facts should be discovered experimentally by models or objects; and thus, in teaching geometry and mechanics, the facts should first be taught experimentally by the use of models and apparatus before the demonstrative proofs are given. Knowledge is thus rendered attractive and pleasurable instead of irksome.

The routine of school hours might be broken by lectures on physics, experimental instruction, and excursions into the fields of Nature. The method of Nature should be closely followed, and vigorous thought will be the certain reward.

These principles lead to what we have before recommended in speaking of drawing. Children prefer colouring pictures to copying lines; more advanced pupils prefer landscapes to drawing geometrical forms. Now the only reasonable course of following this natural bent of mind is to impart such instruction through the instrumentality of *models* and *real objects*, instead of copies, or a dry analysis of elements, which in the case of languages has been condemned. Thus, geometrical definitions of lines, points and angles, should be taught after a sufficient practical knowledge of form in the concrete. In no other way can some pupils acquire this branch of education. Cubes and solids should be presented before planes; the construction of pasteboard models of the solids should be encouraged, as the cube, pyramid, octahedron, &c., in all cases proceeding from simple to complex forms. Thus, by a course of empirical geometry, constructive art of the most useful kind may be learnt by a series of tentative efforts, employing both the intellect and interest of the pupil. It is also better before learning Euclid demonstratively, that the pupil accustom himself to bisect lines and angles, to describe squares and polygons. The same remarks extend to perspective; and trigonometry and conic sections become both entertaining and valuable in exercising the inventive capacity of the young pupil, if pursued in a like manner. Thus habituated to contemplate relationships of form, quantity, weight, and other abstract ideas, the pupil will at once estimate the value of demonstration when he begins, and will regard it as supplementary to the practical problems he has learnt. We are led to the third great principle.

3. *Self-discipline and technical training.*

If the young pupil were earlier taught to investigate and discover for himself as much as possible, there would be fewer mediocrities, and less of that slavish adherence to precedent which hampers often the finest intellect. By the tentative processes we have inculcated, the pupil is led by degrees to discover relationships, and to test the correctness of his eye, and other perceptions. By cutting out paper forms, and forming solids, the scientific conceptions of the pupil become clearer, and evince a desire to substitute something more rational for his ocular guesses; hence he proceeds from indefinite to definite ideas, he progresses from the concrete to the abstract; and eventually attains, by a natural order of induction, that power of generalisation which should be the great end and aim of his school studies and the direction of his professional skill.

Auguste Comte has shown conclusively that the process of historical evolution is precisely that of the development of the faculties of the individual, and the progress of knowledge from the initial to the positive stage of science; in other words, that the *order* and subordination of study should follow the same course as the evolution of mankind, " every science having been evolved out of its corresponding art."

HINTS TO PUPILS DURING THEIR TERM OF ARTICLESHIP.—Our pupil during his three or five years of office training will have occasional opportunities afforded by slackness of business, as well as the leisure hours of his daily work, to devote himself to such studies as he may have neglected or only partially acquired at school. We have pointed out those most essential to his future vocation; they may be summed up as mathematical and general knowledge. He may alternate his reading in these important subjects by acquiring a general knowledge of chemistry, geology, light, acoustics, heat, and other physical laws bearing upon his profession. Fergusson's " Handbook of Architecture," Rickman's, Sharpe's, Scott's and other works on Gothic architecture may also be studied now to advantage. They will enable him to keep pace with the more mechanical routine of his daily work by informing his mind upon the historical development and vicissitudes of his art, and thus making far more interesting the otherwise irksome routine of his labours. Too often the pupil neglects these opportunities of collateral study; he fancies his office work too laborious to devote any of his evenings or leisure to such reading, forgetful that it enhances the interest of his work, and imparts pleasure and attractiveness to what he regards with distaste. His indoor gratification or amusement may also be promoted by procuring a set of tools, and exercising his ingenuity in forming models of a constructive kind, such as frames, roof-trusses; or in making sections of the cone, and other solids, in wood; forming in cardboard the coverings of solids, especially those which are of most interest in domical and roof construction—*e.g.* the hemisphere, and various forms of pendentive, pyramids, truncated solids, &c. The sections produced by planes, cutting cylinders, prisms, and other solids, will enable him to intelligibly understand the principles of vaulting, groining, and domical construction, and the intersections of various forms of roof, far better than a course of reading unaccompanied by such practical illustrations. His mind will get habituated to grasp the geometrical difficulties of sections, and the working drawings of such constructions he may be entrusted with, while the aid afforded will lend a zest to his studies at the office. As a rule there is a sad deficiency of this practical discernment among young men in architects' offices, especially those who, confined in London offices, seldom have opportunities of seeing for themselves actual works in progress; and hence the need for such an alliance of the practical and theoretical.

But his leisure hours may be frequently relieved, especially in the summer months, by rambles out of doors, excursions into the country for the purpose both of pleasure and profit. His sketching-book, pencil, and rule should be his invariable companions. The remains of any ancient building should be examined, and those parts measured and sketched which seem to afford any new idea or mode of construction. Details and mouldings are generally of more value to the student of mediæval remains than their arrangements as a whole. They may be carefully measured and drawn, and notes appended explaining any particular part. The student is particularly recommended to observe *closely* and *practically*, not with the eye of an artist or sketcher only; he should observe and take notes of the material, stone, &c., used; the age and state of preservation of the work; the joints and mode of construction being carefully noted and drawn. Mere sketches and shaded drawings are of little value for reference; free illustrations by *plan* and *section* should invariably accompany the notes. But while he may spend his holidays in these rambles over the tenantless monastery, mediæval church, or ruined abbey, they should not entirely engross his attention; his profession is one of ceaseless progress— he has to meet *modern* not ancient requirements; hence the zeal and ardent enthusiasm of the connoisseur and archæologist should be carefully kept in check. To this end our student should study *modern buildings* of merit more sedulously than ancient shrines; he should take as much pains with and as carefully take notes and sketches of the commonest things, the arrangement of cottages, the fitting and contrivances of joiners' work, framing of roofs, construction of floors of iron and wood, window-sashes, casements, and all details which are of everyday use, and upon which the health and comfort of occupiers depend, as he would in drawing the most elaborate piece of architecture. If he have facilities, as those afforded by the visiting and sketching classes of the Architectural Association, for examining buildings and works in progress, he should be particularly observing, and note all he sees. Details of masonry, plumbing, and ironwork, no less than the finishing trades, should be carefully studied. In all his out-of-door studies let the *useful* take precedence of the ornamental: ornamental accessories may be learnt from books and copies, but constructive art only can be acquired by a course of practical and theoretical training. Schools of art afford ample means to the young student in perfecting his freehand drawing. The excellent lectures of the London University, and other colleges, and the societies of architectural and engineering bodies, afford also the means for acquiring a thorough knowledge of the theoretical principles of architecture and its allied branches.*

We would also recommend the young pupil one very necessary aid to self-improvement for which the majority of architects' offices do not afford the time or means— to make notes of any difficulty or question that arises in the course of each day's work, and to solve or endeavour to master the subject at home. Never be above asking questions of your companions; they, like you, benefit by the inquiry. One word more: keep a *note-book* as well as a sketch-book, and jot down under proper heads every passing thought that occurs and is worth retaining, and every valuable fact or principle you acquire in reading and study;—while it aids the memory, it also serves to give your ideas clearness and precision.]

* The lectures and classes of University or King's College are valuable to the student as preparatory to habits of thought and observation, and enable him to take the full advantages of practice. Such courses of study are also afforded at Liverpool, Manchester, Southampton, and elsewhere, and in the technical classes in connection with architectural societies and provincial schools.

PART II.

STUDIES ABROAD.

OUR quondam school-boy is now, in reality, a "young architect." He has "served his time" in the office of some established professor and practitioner, and we have only to hope that his time has served him. Presuming that it *has* done so, even to a greater degree than is usual, he must still consider, not only that he has much to learn, but something to *un*learn; for the *mannerisms* of his master have, most likely, a present influence upon him to the prevention of the due development of his native taste and feeling. He must consider that he has been hitherto exercised only in those particular styles of the art which his tutor has been called upon to practise, and that he (the pupil) may have, in his future career, to deal with other styles, and even with the same styles in a novel manner. He may have not only new combinations to effect, but also original, or hitherto unrevived, features to study. He has to get the wheels of his mind out of the ruts of habitual office practice, and to drive the coursers of his imagination over the free common ground of varied and speculative design. He has, no doubt, acquired much *artistical* knowledge that is true, and much *practical* attainment that is valuable; but the very conditions of his pupilage have enforced an obedience, which, though most wholesome in respect to discipline, has yet trammelled his invention and checked his fancy. Young architects will be generally found to criticize the works of others by the standard of their master's; and, by the way, they are usually much given to criticism, with a greater aptitude for censure than eulogy—the natural result of limited knowledge bearing upon comprehensive variety.

TRAVEL.—Now, to get rid of the mere bonds of habit, there is, unquestionably, nothing so certainly efficacious as TRAVEL.

> "Home-keeping youth have ever homely wits.
> I rather would entreat thy company
> To see the wonders of the world abroad;
> For it may be impeachment to thy age
> In having known no travel in thy youth.
> Experience is by industry achieved,
> And perfected by the swift course of time;
> Nor can he hope to be a perfect man,
> Not being tried nor tutor'd in the world."

Submissive obedience has been already sufficiently practised. A lively and acute observation has now to be cultivated. What is sterling in the range of former acquirement will not be lost nor diminished. The ever fresh air of changing scenes and differing countries,

> "Puffing at all, winnows the light away,
> And what hath mass, or matter, by itself
> Lies rich in virtue, and unmingled."

Corrective in respect to *past* studies, travel will prove also highly suggestive in regard to the studies which are to follow; and the young artist, instead of remaining a critic over others, will find enough to do in criticizing himself.

Nor let it be supposed that the benefits of travel are less than they formerly were, because books have multiplied to us the labours of former travellers.

The object of travel, it is true, is not so elementary as it was, ere Stuart and Revett, Denon, Taylor, Cresy, and others, had afforded all necessary information as to the details of Greek, Egyptian, and Roman design; but it is, as ever, important in expanding the taste for the beautiful and picturesque, and in stimulating that professional enthusiasm which can only be excited by beholding the actual realities whose distant features we have previously learned to appreciate.

Apart from the more professional and technical matters of a young architect's travel, are others of a moral and social kind not less to be considered. It has been said that "manners make the man," and in no case is the saying worthier than in that of an architect who depends not more on his ability to answer the duties of employment than on the address and conduct necessary to form and secure a connection. Of all men engaged in the polite arts, he is the most frequently and continuously in personal communication with his patron. The sources of conversation which travel affords, and the polish which it may be reasonably expected to occasion, are obviously of no mean value to one who may be constantly the table guest and resident visitor of his employer. An accomplished architect is necessarily a man competent to talk at least, if not to evince in some measure a practical attainment, of Art in general. A feeling for elegant literature is also a natural concomitant of the critical refinement which his reading should have secured to him. The knowledge of the continental languages will not have been acquired without some acquaintance with the leading authors who have employed them; and an experience of continental society will not have been effected without an improvement in his behaviour. Many are the instances of young men having formed those intimacies among their own countrymen abroad, which have subsequently proved most productive at home; and certain it is, that he who has enriched his portfolio with evidences of his industry in Rome, Florence, and Venice, will find an advantage in its mere possession as a credential, though otherwise it may serve him but little.

[The studies of the architectural tourist whose means are limited may now be more profitably directed to the northern side of the Alps. The exhaustive works and photographs published illustrative of Roman art; and the doubtful value of Italian Gothic, which has recently received an undue impetus, and has been so profusely illustrated, render a trans-alpine study of less importance to the student who would confine himself to the more direct and practical advantages of travel.]

The truth is, there is no longer any occasion for him to risk his neck in clambering the arcades of the Coliseum, or to spend his time in measuring the portico of the Pantheon. So far, at least, as it regards the details of Egyptian, Greek, Roman, Gothic, Moorish, and Byzantine architecture, his work is already done for him. If he cannot possess himself of the books themselves, he may have ready access to libraries in which every important feature of these varieties of design is elaborately and truthfully delineated. It is his sketch and note book, rather than his measuring-rod, which should occupy his foremost attention. He requires less to fill his paper with dimensions than his mind with *ideas*. He now wants feelings rather than facts [it would be juster to say the student wants *principles* rather than facts]; correctives rather than corroborations; motives rather than materials; speculative freedom rather than academical precision. This is the time for him to cultivate the poetry of his art, ever attentive to those high and catholic principles of design, which, though the same in essence, develop themselves in different forms suitable to the climate, the manners, the religious or social state of the different countries through which he successively passes. He will by no means confine himself, as was the case formerly, to *antiquity*. He will take observant cognizance of the numerous illustrations of mediæval modification; and still more of all examples of more modern excellence. In two instances only will he remain exclusive in his devotion ; viz., to *ancient sculpture* and the *old masters of historical art*. Let him remember, that Architecture raises the temple which Painting and Sculpture are to occupy as their own loved home ; and that, as he may have to co-operate with the painter and sculptor in the production of " one entire and perfect " work, it is a duty he owes to his fellow-labourers to cultivate an adequate feeling for their respective portions of it. He alone, who is in some degree a painter and sculptor (*i.e.* critically), can be competent to the honour of their copartnership. If the young architect be inclined to carry it further than criticism, the period of his travel is the time for his operations. Then may he well vary his pursuits with drawings from the antique and with sketches from the grand frescoes of Raphael and Buonarotti ; but, especially, with exercises in water colour from Italy's own Nature, in her combinations with architectural forms. Highly advantageous is it for every architect to become a correct and ready sketcher, a master of eye-perspective, and a creditable performer with his brush and colours. The fascinations of smart and lightly managed effects of sun, shadow, and tint, will some day " tell " in his favour; and he may now be engaged in preparing for his future drawing-room pictorial decorations, which shall also be of important service to him as so many official insignia, "flags and signs" of the love he bears to the profession he has adopted.

[Though our author's advice was well intended at a time when academical precision absorbed so much of the young architect's mind, it is less needful at the present day when the various and practical demands made upon the profession necessitate on the young practitioner's part some abatement of his taste for the antique and merely archæological.]

His more practical drawing will be well applied to choice selections from the architectural fragments which may excite his admiration in the several great Italian museums, all of which are prodigal in the exhibition of decorative art. The experience already acquired at home will teach him where such things may be hereafter suitable for application; and his employer will not be the less pleased on learning that the vase on his balustrade or the frieze in his chimney-piece are facsimiles of some valued importation from the "Museo Vaticano."

ITALIAN GOTHIC.—Italian Gothic he will carefully eschew—at least as a model. [Since these "Hints" have been published, Italian Gothic has received an amount of popular adoration which would have justly shocked both our author and his contemporaries.] To the great cathedrals of Germany, France, and Normandy his *continental* Gothic studies will be confined; nor will he forget, even in perusing them, that England is, after all, more especially the school in which Gothic architecture develops itself with the most essential truth. In Normandy, the Norman Gothic is unquestionably better and more fully illustrated than with us; and in many of the foreign pointed examples he will see certain individual parts of a far greater magnitude and more elaborate richness than any he can meet with at home; but it is still from an untiring study of the cathedrals, churches, and old mansions of England, that the true principles of Gothic design, the laws of its proportion, and the most effective results of its combinations, are to be deduced.

The growing feeling in our country for the palatial style of mediæval Rome and of Venice, and for the villa of modern Italy, will, of course, direct him to give more than common attention to such examples as best exhibit them; so that he may co-operate with his numerous improving contemporaries in working out a worthy Anglo-Italian school of design. Scientific and literary professors, travellers, High Church conservatives, and others, have all built their club-houses in pursuance of the aim started by the Buonarottis and Palladios. The Palladian palace of Stowe, and the grand piles of Blenheim and Castle Howard, still maintain their ascendancy over all modern attempts at the castellated or Tudor mansion. [The castellated and Tudor styles here alluded to have given place to a less pretentious and extravagant species of domestic Gothic. Here, as in ecclesiastical Gothic, however, there is a tendency to mimic literally peculiarities and quaintness of feature and to affect mere "picturesqueness," a common vice of the ultra-Gothic school which should be steadfastly resisted]; and, while the Church Architectural Societies are effecting much good in the restoration of a pure and correct taste for Christian Pointed Architecture as applied to churches and other buildings ecclesiastically connected, there can be little doubt of the propagation and continued durability of a reviving love for the modifications of Greek and Roman design. [The hope here expressed has, at least in the direction of secular buildings, been realised to a greater extent of late years, since the present fever of Mediæval ultraism in taste has abated.]*

TIME OCCUPIED IN TRAVEL.—As to the time which should be occupied in travel, two years should be the utmost; while one, employed with devotional industry, may be sufficient. [Six months, if studiously devoted, will now suffice.] At all events, a longer period than the former may too much interfere with the business habits of a young architect who only has his profession to depend upon. The writer of these "Hints" was limited in time because limited in means. Impressed with the fear of debt, and anxious to relieve those by whose kind aid he was advantaged, his "travel's history" scarcely filled the twelvemonth. The cost of his travelling, lodging, and other incidentals, did not exceed ninety-two pounds, about twenty more having been expended in books and other articles of professional utility. To him the pleasures of society (save those he enjoyed at the common mess-table of his brother-artists) were denied. Excursions of relaxation and mere enjoyment were out of the question. He witnessed one opera at Milan, because it was his duty to inspect the grand Scala theatre; and made pleasure and profit tell together in seeing at once the interior of a French theatre and the acting of Talma. But he feared the expense of venturing south of Rome; forfeited the desired gratification of seeing Vesuvius and the disentombed cities of its vicinity, the gay beauties of Naples, and the solitary grandeur of Pæstum; and, after all, returned home with as much preserved cash as would have enabled him to accomplish what he had not dared to attempt.

* Since this note was added, a reaction of taste towards a more Renaissance feeling has taken place—a species of "Queen Anne" style, which has affected both buildings and furniture.

[HINTS TO ARCHITECTURAL TOURISTS.

COMPARATIVE VALUE OF TRAVEL.—Since the time of our author that very necessary finish or supplement to the education of the young architect, "travel," has to a great extent at least been superseded. The numerous sketches and drawings which have during the last thirty years been published upon almost every conceivable branch and style of foreign architecture, particularly the Gothic or Mediæval examples of Italy, Germany, and France, and the facilities afforded to the student to acquire such additions to his library; the casts and examples of the various art exhibitions and architectural societies, and the easy reference to such works, have considerably lessened the necessity there once existed for the young architect to personally visit and examine the remains of ancient art. Not only have architectural works brought home to the student the most famous of continental and classical examples; but the art of photography has with Nature's own pencil revealed with unerring truth the beauties of detail of every object upon which diligent artists and *dilettanti* have devoted their whole lives to depict. Yet we advise those who have means and time at their disposal, either before their entrance upon practice, or occasionally during those holidays which custom has especially set aside for tour-making, to avail themselves of those pleasant rambles abroad which serve to enlarge the scope of their ideas, and bring home to their daily vocation fresh vigour and more catholic views of society and art.

ITALY.—Rome, Florence, and Venice, although hackneyed resorts of the architectural tourist, may be visited with profit. Comparatively exhausted as Italy may be, there yet remains ample fields from which an unbiassed artist may glean some materials left untouched by those who have tracked them before. But there are other lands besides " fair Italy " to attract him. He may gather much from more northern routes; the Spanish peninsula, and even the capitals of Russia, comparatively untouched ground, may afford him an unexplored field for his researches. To profit by tour-making, the young tourist should become acquainted by previous reading with the localities he is about to visit, both historically and topographically. Their geological and other peculiarities should be known beforehand. Such preparatory knowledge, so far from detracting, materially aids the mind to an independent opinion and estimate of the place and buildings visited; it enhances the interest of actual sensible gratification.

One word to the architectural tourist may save him a regretful retrospect upon his return home. Provide yourself with both a sketch-book and note-book, and let every excursion and ramble be not only fully described but freely commented upon. Let the pen as well as the pencil speak. One or two other hints should be carefully kept in view by young tourists.

If you go to any church or ruined abbey, let not the mere sentimental side of your convictions overbalance your stronger reason or the higher duties of your calling. Let not the devotion to a shrine, or piece of mosaic work, however lovely in your eyes, or a bit of sculpture, carving, or embroidery or metal work, absorb too much of your attention, or wean the mind from any of those constructive peculiarities of the work which are worth noticing.

Try to discover the cause of any beautiful effect in construction and design. If proportions strike you, measure them carefully, and study them on paper. Take note of any particular mode of construction, and do not rest content with a general outline or sketch. Constantly keep before your mind the fact that the most beautiful and stirring work of architecture is the result of *constructed* art, arising out of necessity and thought.

Another important point should be kept in view. The fascination of travel, the associations of an emotional kind awakened when contemplating scenes rendered classical by literature and song, often with bewitching enticement, induce a blind infatuation for everything foreign from which the young enthusiast seldom recovers. The writings of Ruskin on Venetian, Veronese, and Florentine art have already captivated the popular mind, and led, siren-like, some of our "fashionable" architects to reject their calmer reason and to reproduce styles absurdly out of place in this northern climate. Let the student beware of this allurement, and while admiring the beauties of the South, the land of classic art, of painting, mosaics, sculpture and marble, remember that he is working *in England*, not across the Alps, and that the benefit he derives from such works will be in proportion to the amount of original thought he bestows upon them, and the means they afford of enabling him to grasp the more completely the immutable principles of his art. A wise eclecticism in art is preferable to wholesale copyism, though the stock of ideas acquired by travel should enable its possessor to generalise and perfect more completely the elements of art. What he sees should not be considered as giving him principles or methods, but simply certain *results*, the products of particular social and material conditions. A critical mind should be brought to their investigation. Let not peculiarities be mistaken for beauties, nor the imagination run wild in drawing poetical analogies between fact and sentiment.

FRANCE.—The architectural student whose means are limited cannot take a more pleasant ramble than one into Normandy. Its contiguity to England, the facilities of conveyance, and the valuable and prolific examples it contains of the continental development of Pointed architecture, from the simple beauty and expression of the early Romanesque, through the various phases of the style to the transition into the Renaissance, make it a favourite tour with the young architect. Though our own country is rich in the Romanesque or Norman style and the transitional Early Pointed, we have not the means of studying those earlier germs and modifications of Gothic which especially adapt themselves to the requirements of ordinary buildings in this style. The forms of the Early Pointed churches are well worth study for simplicity and purity. French Pointed work in the north is more nearly like our own, as at Lisieux, St. Ouen, while the introduction of the classic forms are less prejudicial than they are in more southern parts of France. Rouen, dirty as it is, demands the attention of the student, especially its cathedral, where the history of the Pointed style can be traced. The Palais de Justice and St. Maclou exhibit the admixture of Gothic with Renaissance forms. St. Ouen shows also the florid character of the fourteenth and fifteenth centuries, its proportions being second to none in France, while the choir and transepts are perfect of their kind, both in design and richness. Mr. Fergusson thinks it the most beautiful church in Europe, and it certainly is one of the purest of its style. St. Maclou displays the gorgeous and elaborate detail of a less pure Flamboyant type, where the student may learn to discriminate between the elegance and defects of fifteenth-century French Gothic, which cannot compete with our own florid specimens of the style,—Henry VII.'s chapel to wit. St. Jacques at Dieppe is another work of the class which the student will do well to avoid, though he cannot fail to be struck with its overwrought detail.

For earlier examples, St. Stephen's, St. Nicolas, Caen, will repay study. Coutances, Bayeux, Honfleur, Caudebec, Jumièges, Bocherville, may be visited *en route*.

If the student can afford time, a tour through the southern provinces of France will give him an idea of the mixture of pure Frankish and the Romanesque or Romance styles. In Provence he will take cognisance of the pure Romanesque or Round-Arched Gothic, verging indeed to classical Roman; such is the cathedral at Avignon; or mixed with Pointed features, as the cathedral at Vienne. The circular form of church found here is worthy the student's attention, as also the mode of vaulting seen at the church at Fontifroide, near Narbonne. Aquitaine possesses many examples of a style partaking both of the round-arched, and a domical pointed style of an Eastern character, well worth study. The *internal* buttressing of the walls, as at Alby, is suggestive to the architect; so also the cathedrals at Bordeaux and Toulouse, with their grand aisleless naves. The Round Gothic of Auvergne exhibits a pure style which Fergusson rightly regards among the perfected styles of Europe.

But whichever part of France the student's rambles may take, his attention should be directed especially to the magnificent development of thirteenth-century examples as exhibited in some of her cathedrals. The thirteenth century was to Gothic art what the Periclean and Augustan were to Greek and Roman. The cathedrals of Chartres, Rheims, Amiens, Coutances, Beauvais, Troyes, Noyon, Paris, are, in the order here given, especially deserving study. One other, the purest and finest after the first three mentioned, namely, the cathedral of Bourges, must not be forgotten. Its proportions, its aisles, its symmetrical plan, its chevet, make it a model of French Gothic.*

For further remarks on " style " in design the reader is referred to Section 3, Part VI. The student who desires to avail himself of the labours of brother tourists in this field is recommended to study the works of Messrs. Norman Shaw, Nesfield, and the beautiful drawings of Johnson's " Early French Architecture."

The brick architecture to be found in Italy, Germany, Spain, and France is particularly worthy of the attention of the architectural tourist, and Mr. Street's elaborate " Brick and Marble Architecture of North Italy " may be read. (See also Ruskin's works.)

While we thus refer to Continental studies, let not the young architect neglect his own country. England possesses by far the purest examples of Middle-Age architecture; and we recommend the student to study well the principles of development and design observable in the thirteenth and earlier fourteenth century buildings. For purity of outline and form, in vaulting, pillar-clustering, and mouldings, no foreign examples can compete with the English. The student is recommended to study, 1st, *Masses* or *Forms*; 2nd, *Lines*; and to observe that in the best periods of art both had equal attention bestowed upon them. A careful study of Mr. Edmund Sharpe's admirable "Architectural Parallels," and his work on the "Seven Periods of English Gothic," is recommended. Many of the Cistercian monasteries of the twelfth and thirteenth centuries are excellent models for study.]

---

* The student of French Gothic should note especially the development of vaulting, as seen in the earlier examples.

PART III.

EARLY PRACTICE.

OUR traveller has now returned. His brass plate is upon his door. He has indentures to prove his apprenticeship, a portfolio to assert his subsequently acquired accomplishment; and he is ready to begin.

The probability is that he'll have to wait awhile. He will have nothing to do,—or what he does will be done for nothing. Some one will kindly give him an opportunity of showing what he *can* do, the favour shown and the labour given being mutually gratuitous. Advertisements will invite him to compete for a Town Hall, or a "New Bridewell," a Market House, or a New Poor Union; and he will send his plans forward, and they will be sent back; and some one already well to do in his profession will, as he is informed, either by favour, or job, or otherwise, win the premium and be commissioned to carry on the work: and thus, with the rejected among many, sit down disconsolate, and quote from Jaques—

> "Thou mak'st thy testament as worldlings do,
> Giving thy sum of more to that which had too much."

And then will he be stimulated by a promise from some worthy friend of his father, who expresses vague ideas of "some day adding a new dining-room to his house," under the inspiration of which, visions of a sideboard recess, flanked by Corinthian columns, suggest themselves; and lastly, he who has promised nothing shows his friendly indignation in abusing him whose promise has turned out to be nothing worth.

Hopes, disappointments, and efforts (for the present) unavailing, will (unless he be wondrously fortunate in chance or connection), mark his career for some time at least; *mais le bon temps viendra*, and we propose filling up the leisure of the interval by putting before him such matters for consideration as may make him rather value than otherwise the spare time which yet lies upon his hands.

The duty he now owes to himself is twofold. In the first place, he has to form and increase his connection by constantly availing himself of every opportunity for manifesting his professional claims to desert. In the second place, he has to prepare himself for an effective and perfect fulfilment of the duties which his first engagement will impose upon him. We have already sought to impress upon him the heavy responsibility which will be his when he is no longer the mere agent of a professional superior. Let him not postpone this reflection until the day of employment arrive. Everybody is always in a hurry to have everything done. His patron will take six months to think of what he desires to have accomplished in as many hours. When the commission arrives, immediate work will be required—not preparatory study; and if there be not a ready foresight to pierce through all contingencies, the progressive and ultimate perplexity will be proportionally bewildering. To anticipate possible objections is greater policy in an architect than to give immediate answer to requirement. Of all professions, his is the one most subjecting its professor to meddling interference, and a thoughtless disregard of trouble taken and obedience unrequited.

> "Double, double,
> Toil and trouble,"

is indeed the chant of the sister Fates who are hostile to an architect's peace. The graces of the portico, the beauties of decoration and proportion, the triumph over a hundred contending desiderata, shall be all forgotten in my lady's passion for — a housemaid's closet! It availeth not as an excuse that you can put it under the back stairs. "It should have been thought upon before. An architect! and not think of a housemaid's closet! It ought *not* to be an extra." "Extra!" Fearful word! The builder's aim, and the architect's dread! Let our young friend think of it betimes; and let him bear in mind, that the best guard against the overwhelming censure which follows it, is to habituate the mind to a foresight, which, during the study of the nearest and most important things, should penetrate into the most remote and trifling. All the grand principles of design, convenience, and enduring strength, may have been perfectly answered by the most artistical ability, by ingenious arrangement, and constructive skill; but if chimneys smoke, gutters leak, or drains choke; if windows prove not in all trials weather-tight; if all the little conveniences of the former house be not added to all the larger ones of the present; if a shelf, a cupboard, or a rail and pins be omitted where custom might expect to find them; if the whims of old servants be not considered, or the carelessness of new ones anticipated; — if, in short, the genius of a Michael Angelo be not followed close up with the care of a cabinet-maker, the architect will yet have a toil of vexation to encounter which may make him almost repent the choice of his profession.

Practical Hints.

Hints on Planning.—We shall begin our practical hints with some remarks in reference to plans, or internal arrangement, as affecting elevations, roofs, and chimneys.

The young architect too frequently concentrates his attention on those portions of his plan which concern one or more particular façades. Thus, he is careful of his entrance front and his lawn elevation, as those alone which will be visible to a stranger approaching from the lodge, or walking in front of the sitting-room windows; and no sooner is the building roofed in than he discovers that the "return fronts" are provokingly more generally visible to the public eye from without the boundary of the premises than the others which have had his too exclusive care. One of his "architectural" elevations is seen in continuous connection with a surface of unstudied masonry, the respective parts of which neither harmonise in position nor in decoration: or, at the best, he exhibits a display of blank architecture, the falseness of which is proved by certain prominent necessities which will not be either concealed or modified. The offices and other inferior appendages to the mansion cling to it, and proclaim themselves with all the humiliating impertinence (or rather *pertinence*) of poor relations bent on the declaration of their consanguinity. The idea of "planting them out," which originally existed in the mind of the designer, still exists in *his* mind only. The trees he requires will take at least fifty years to grow; and, even then, winter will in its turn disrobe them of their foliage to leave displayed an obstinate range of architectural poverty.

Evergreens will *never* grow high enough. The whole thing must remain as it is—a handsome countenance with an ugly profile:—a beggar in a velvet waistcoat, and no coat to cover his sides.

This oversight is still more commonly committed in town houses and street architecture. Nothing is more frequent among builders and young architects than the exhibition of a mere mask, which only deceives while the spectator is directly opposite on the other side of the street, or so far as there may be houses of equal height continuing on either hand. Otherwise, directly the front is passed, the blank masonry or naked gables of the returns show themselves like the mere party-walls in the transverse section of an unfinished range; and these, be it remembered, are often seen for a much greater length of time than is given in passing, to the main front, since we may have them before us during the whole of our progress along a street of half a mile extent. Perhaps only a portion of the return ends may be seen above the roofs of the lower houses adjoining: but it is not the less necessary to continue along this portion the architectural character of the front. In the many instances which occur of houses rising successively one above the other on the side of an ascending street, too much care cannot be taken to give a finished perspective effect. The means will readily suggest themselves. Architecture, as we have before said in the first section of our "Hints," has a peculiar privilege among the arts in commanding observation from the distance, and no town or range of buildings will ever have an imposing or even a tidy appearance, while it shows itself to be composed of independent fragments jostling one against another. The beggarly habit of carrying a cornice or parapet, with dressed doors, windows, pilasters, &c., along a twenty-feet front, leaving in barn-like nakedness a thirty or forty-feet end, is an abomination which even the most vulgar country builder should eschew. Infinitely better that the whole should be consistent in the absolute perfection of nudity.

REFERENCE OF PLAN TO ELEVATION.—Oversights, however, sufficiently unpardonable are often exhibited by architects of more established repute, in settling their *plans* without due regard to the final appearance of their exteriors. [Among some leading Gothicists this is painfully evident. Want of balance, unequal distribution of windows, awkward intersections of roofs and gables, and other irregularities are the result. In designing a plan the block outline of the ground, if confined, should be carefully studied and kept in view. The position of the entrance and principal apartments as regards aspect or front should then be considered and laid down as the key-note of the plan; the distribution of the other rooms and offices should be made to agree with such an outline, at the same time the outline or disposition of the main masses of the building should be constantly in view, as regards the roofing and elevations. In fact, experience has shown us that the *ground plan, principal elevations,* and *roof plan* must be simultaneously studied to produce a satisfactory whole. The natural order of arrangement appears to be:—1. Position of entrance. 2. Communication. 3. Aspect of apartments. As regards the elevations, the roof plan must first be considered in disposing of the projections, gables, and other features after the general arrangement has been hit upon, and in complex buildings the cross section should also be drawn.]

ARRANGEMENT SHOULD SUGGEST STYLE.—Let it, then, be the first care of the young architect in designing his *plan*, to do it with especial reference to the [*fitness* or *purpose* of his structure, disregarding the mere "style" of architecture]. In strictness the style should be suggested by the internal arrangement; but, either way, it is equally an architect's duty to see that convenience and external expression be true to one another. A ground plan may be exactly adapted (by certain equally convenient differences of arrangement) either to a Greek or Roman, a Gothic or Italian elevation; and whichsoever of these may be decided on, the arrangements of the walls, with their breaks, recesses, and projections, and the position of the fireplaces, must be thought of in close conjunction with the ranges and intersections of the roof, and the satisfactory position of the chimneys, as objects in the general view of the building. It is not, in fact, until a plan of the roof is made, with its stacks of flues well located and accurately drawn, that the masonry of the floors beneath can be decided on: nor should the plan of any one floor be *finished* until those of the floor or floors above are perfected.

These points being all considered, the young architect will take care to make his elevations honestly exhibit their crowning roof and chimneys. The custom of omitting these features is seriously reprehensible, and worthy only of a Pecksniff.

ADVANTAGES OF PERSPECTIVE SKETCHING.—He will be equally careful also to show *all* the fronts, and to give at least such *perspective sketches* as may prevent those common misconceptions which *geometrical elevations* occasion in not truly showing the projecting or receding of the different portions. Even architects deceive themselves by the pleasing effect of façades geometrically developed; an effect which is rarely seen in reality, except at such a distance as renders indistinct all the decorative details of the building. What a false idea, for instance, does the geometrical

figure A give of the perspective figure B! It is not enough to show, by the plans or by description, that *c* projects and that *d d* recede. The strict truth is, that, in the perspective view most generally visible, the building will lose all the expression of *length* which pleases in the geometrical elevation, and will become a *short* squat building, with only one visible wing instead of two.

[Façades or elevations of buildings in which prominent projections do not occur may be sufficiently represented by shadows projected at an angle of 45°. By this means all slight projections vertical and horizontal can be at once discerned, and their real projection also obtained on elevation.]

The geometrical elevation of a circular temple is most deceptive in its appearance, and will occasion expectations of much greater width than a near perspective view will exhibit: thus, the building which will show geometrically as fig. 1, will show perspectively as fig. 2. In the elevation of the west front of St. Paul's Cathedral, the tambour of the dome looks overwhelmingly large; in the view of the real building from a moderate distance it exhibits no such excess.

Fig. 1. Fig. 2.

VALUE OF ANGULAR PERSPECTIVE.—Again, the geometrical elevation of a square tower, in which the expression of great altitude is required, should be made with reference to the increased bulk it will exhibit, when viewed diagonally, to the prejudice of its loftiness. The habit, in short, of considering the elevations of parallel planes, without equal regard to their "returns," and without studying the diagonal view of both united, is the cause of infinite disappointment; and equally so is that of only looking point-blank against edges and *vertical* surfaces, without duly reflecting on the additional effect of under, or upper, *horizontal* surfaces. For example; what a light and simple effect has the common cantilever cornice, fig. A, compared to the same cornice seen in perspective, as fig. B. Features, which in the geometrical drawing may appear light and well-proportioned, may in execution prove ill-proportioned and heavy. Again, what may seem well developed in the drawing, may wholly or partially disappear in the work itself, as in the case of a parapet or blocking course concealed from the near view by the projection of the cornice.

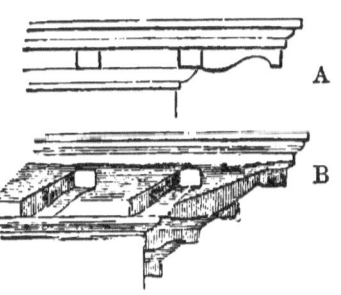

A dome which, geometrically, has a sufficient height, may, from the point of most frequent view, seem offensively flat. Sir C. Wren, aware of this, has formed the outline of St. Paul's dome by segments of circles struck from two centres like a Gothic arch, the point of meeting being concealed by the base of the lantern. Its *appearance*, however, is that of a perfect semisphere.

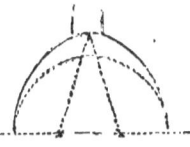

There is also

another precaution to be always carefully taken in the management of circular buildings, and this refers to   the unpleasant effect of overhanging segmental architraves or soffits. A is the elevation of a window in a bow projection. B is its perspective appearance from one side. This is not less objectionable in respect to its constructive weakness than in regard to its ugliness; for it is only by concealed management that a flat soffit arch on a curved plan can be made to stand at all. The case is still worse when the window-head is a curve, and, in short, this practice is only allowable when the curve of the plan is so large, and the openings so narrow, as not to leave perceptible the defect of the overhanging segment.* Thus, in the vast curved outline of the Coliseum the arched colonnade is unobjectionable. In the closely set peristyles of St. Paul's dome, and of the Temples of Vesta at Rome and Tivoli, it is equally so; but where the curve of the plan is small, and the openings, or spaces between the columns, proportionally large, it is a grievous fault. Where small 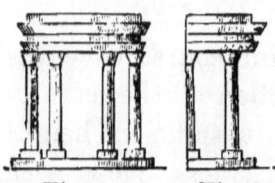 bay projections are desired, they should always be semi-hexagons or semi-octagons, with the windows in the flat faces, unless indeed the required bow window may be so subdivided by mullions or pilasters as to remedy the objections stated. The semicircular portico, fig. 1, may be sufficiently pleasing in its front view; but a glance at fig. 2 will

Fig. 1.   Fig. 2.

[* For similar reasons, circular corners to buildings of small radius are very objectionable, and arches constructed in them defective in stability. Such corners should be canted.]

show the necessity of studying, not fronts only, but profiles also.

Triangular Plans.—While on the subject of the different appearance of objects in different points of view, it may be as well to refer to the triangle as a form of plan frequently, and most injudiciously, adopted in pyramids, obelisks, and pedestals. Viewed directly in front, on the lines *a—b* or *c—d*, it is well enough, as shown by figs. 1 and 2; but who that sees its appearance on the line *e—a*, as shown by fig. 3, does not at once observe that no pyramid or obelisk should ever have an odd number of sides? For the same reason, tripod pedestals should be most cautiously used; for, whatever may be said in favour of fig. 4, it is obvious nothing can be adduced in defence of so ill-balanced a composition as fig. 5.

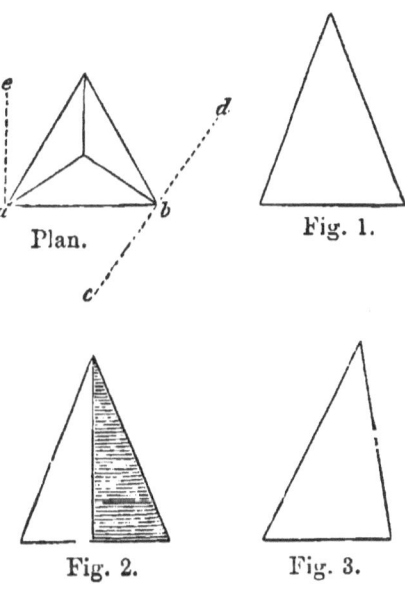

Plan. Fig. 1.

Fig. 2. Fig. 3.

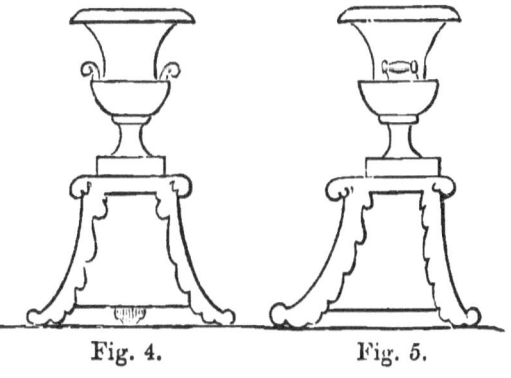

Fig. 4. Fig. 5.

All this goes to prove the necessity—if not absolutely of models—of the perspective effects which buildings will have from all points of view. The architect, pleased with his front elevation, may find reason to alter it the moment he turns the corner.

Solids and Voids.—A different proportion in the solids and voids of a main and return elevation will be fatal to good effect. A material difference in the distances between the common angle and the windows of the front and return façades, or a much more crowded position of windows in the one than the other, will be offensive; and it may be here remarked, that the proportions of solid and void which hold good in the case of a simple façade with no dressings to its doors and windows, will not equally serve when those dressings are to be supplied, since it is only the *plain* part of the pier, or of the space between the lower and upper apertures, which will "tell" in the matter of *breadth*.

Window Dressings.—The architraves and flanking columns of a window must be regarded as the window itself; and as a general rule, it may be said that there cannot be an adequate expression of breadth, unless the plain part of the pier be equal to the entire width of the window and its dressings united. The same law holds good in the horizontal spaces, which should exhibit, in a large and ornate building, the same amount of plain masonry above and below the architraves, strings, cornices, &c., which, in a small and plain building, would intervene between the sills and soffits of the windows and the strings or cornice below or above them. This, it may be remarked, amounts to little less than saying that architectural decoration is more applicable to large than small buildings, and it is true; for doors and windows do not increase in the same ratio that the size of the building increases. On the contrary, they generally bear a much greater relative proportion in small houses than in princes' palaces; and assuredly, where they do not leave, *at the very least,* such a breadth of pier as will allow the width of the opening to intervene between the dressings of two adjacent windows, the latter had better be left without dressings. On no occasion whatever ought the breadth of a pier to be less than the width of a window opening; *i.e.* of course supposing the window to be a single one, and not triple with intervening mullions.

JUNCTION OF MAIN HOUSE AND OFFICES.—If the error of making a discordant difference between the fronts of the main house be so serious, not less so is the total discordancy often seen between the main house and the offices. Now it is, in fact, very rarely that the offices are not, from several important points of view, seen in conjunction with the principal mass of the structure; and the difference therefore between the two should strictly be one of degree only. A handsome cornice along the eaves of the one will be ill-accompanied by a common eaves gutter along those of the other. Correctly speaking, it should be of the same form, reduced proportionally in scale, and—if required—without the enrichments of the main cornice. Above all things, the young architect should avoid the common mistake of reducing the beauty of the chimneys; for those of an office range, springing usually from a lower roof and having a greater relative altitude, will very likely be more conspicuous than the others. In short, an aptitude for chimney design is most important to an architect engaged in villa building. Let not the anticipation of chimney-pots escape his consideration. On the contrary, let him design them, and show them in his elevations, as likelihoods, which, if ultimately necessary, may not be absolutely disfiguring. He will further remember, that, where the flues in any one stack are numerous, it may be better to place them in united parallels than in one continuous range; and he will be also cautious in so arranging his fire-places as that the various flue-stacks may be as nearly as possible of one size. This uniformity, at all events, should be observed in corresponding pairs of stacks.

CHIMNEYS.—He need not be reminded, that, in Gothic structures, chimneys are not only admissible, but are often advantageous in their attachment to outer walls—especially when they rise with the gables. In the free and irregular style of the Italian villa they may also occasionally be connected with the outer walls. In the severer Roman style they may rise from the angles (as shown in Barry's Reform Club House); but in no style (saving only the Gothic) should they rise from the eaves if it can possibly be avoided.* The inordinate height required to raise them above the ridge of the roof, their insecurity (involving often the application of iron struts to sustain them), the difficulty of a satisfactory management of the main cornice beneath them, and the plumbing required to make weather-tight their union with the slates;—all these circumstances make it most desirable the plans should be so arranged that the chimneys may ride, as it were, upon the ridges of the roof.

Flues.—The occasional practice of making flues run a long raking course in the thickness of walls, and of

[* These observations and restrictions as regards chimney-stacks are correct as far as they agree with the conventional rules of taste laid down by architects, though the young practitioner is advised to exercise his own common sense and discretion in designing such important functions of a building untrammelled by the mere pedantry of academic rule. Stacks should be collected as much as possible, and placed in positions whence they can rise uninterruptedly, without sudden bends, through the roof. Chimneys attached to walls, and projecting therefrom, often relieve a blank front.]

making them even turn corners to conduct them to a desirable position of exit, cannot be too seriously reprehended. Underground flues, too, which must be periodically opened to be cleaned, should never be adopted save under those imperative circumstances which the most industrious ingenuity cannot avoid. Never allow two flues to unite in becoming one; and above all things, so arrange the floor and roof timbers, that there shall be no chance of their being carried (even by carelessness itself) into the flues, or within at least nine inches of them. Here let us remark, while the occasion so seriously calls for it, on the necessity of an architect never trusting to the *sagacity* of workmen —especially in the country. The common carpenter and rubble mason will each do *his* work irrespective of the other's; and, on visiting your building, you will very likely find that a joist or a purlin has little to divide it from the fury of a chimney on fire, except the plaster pargetting which lines the inside of the flue! With equal care, look to the work which receives the hearths of the fire-places. "Brick trimmer arches" may have been inserted in the specification; but if there be not a clerk of the works to look after the building, it is by no means certain they will be constructed.

CONNECTING LINES IN COMPOSITION.—To recur to the subject of the offices and inferior buildings attached to the main structure. It may require some care to make a good junction between the lower roof of the former with the higher one of the latter, unless the ridge of the one can be brought under the cornice or eaves of the other. Again, the union of the main and inferior structures should be so harmonized, by the use of certain string-courses or lines, common to both, as to show that the two or more parts are component features of one whole, the extension of which is not so much that of connection as continuity. Finally, it will be well to avoid the probability of future appended additions, as out-houses, lean-to's, &c.; and never to put off the consideration of wood and coal houses, shoe and knife houses, dustholes and privies, until the mass of the building is up.

POSITION OF WATER-CLOSETS.—There is one particular necessity in every good house which the young architect should consider from the very first. It is a necessity not often very successfully met, because it is rarely considered until too late for efficient management. Water-closets are generally so placed as to make it a very difficult matter for ladies and gentlemen to conceal from one another the more humiliating circumstances of their common nature. It may be, that we are in England too nice on this point; but it is nevertheless a point on which an architect is privileged to exert his ingenuity. So place a water-closet that any one going in its direction, or returning, is not necessarily going to or coming from *it*. Secondly, let it be so located that its door cannot be seen from the hall, the staircase, or any important part of the house where the inmates are likely to be passing. Avoid, under any circumstances, putting it at the end of a long passage. Be still more particular in so placing the closet that the operations of the occupant and the apparatus shall not audibly announce themselves to the sitters in the day-rooms or the sleepers in the bed-chambers. But, above all, let not the vicinity of the closet be made known by any offence to the nose of nice gentility. [One of the most desirable positions for a water-closet is on a landing, recessed, if possible, from the staircase. It there answers, in small houses, all the purposes of day and night use; and if attached to a sink and lavatory as an isolating chamber, it becomes a highly advantageous adjunct, adding materially to the comfort of a house. It may also frequently be placed at the side of the entrance accompanied by the same conveniences. As to the avoidance of noxious effluvia, the reader is referred to Part V., Section 3, *House Drainage.*] On this last most important matter we would call particular attention to the following precautions. Avoid placing the cesspit within the building. See to the certain efficacy of the stench-trap at the foot of the soil-pipe. Let not the waste-pipe from the cistern enter the cesspit without the water first passing through a trap which shall prevent the ascent of effluvia into the locality of the cistern.* If the soil-pipe can be carried outside the wall, let it be so; only take every care to prevent any of the pipes from being injured by frost.† Let not the cistern be uncovered to the air, lest the mouths of the pipes, &c., be choked with leaves or other matter carried there by wind or rain.

PLUMBERS' WORK.—While on the subject of plumbers' work let us impress upon our young practitioner its foremost importance, and the necessity of giving the fullest description of it. It is usually too much *generalised*. Always employ rolls and drips to unite the sheets of lead. Do not use solder where it can be avoided. Leave the lead free to expand and contract. Cover the gutters with raised boarding. Provide capacious receiving-boxes at the heads of water-pipes. Take every precaution to prevent leaves and rubbish from collecting in the boxes and choking the pipes, and be equally careful to carry the water from the feet of the pipes immediately clear of the building and into the drains, otherwise they will only prove most injurious to the very stability of the building by soaking the foundations.

* See page 201.
† Soil-pipes should be encased either in brick or wood.

DRAINS.—In constructing drains have a close regard to the facility of cleansing them, especially where they unite with the pipes; [at all junctions capped pipes should be used,] and be equally careful in the supply of traps to prevent the progress of rats and vermin. It is too commonly supposed that workmen will have an eye to all these obvious necessities, but workmen might as well imagine that their tools will work of their own accord as might architects conceive that workmen are to be trusted without accurate description and scrutinising supervision. We thus mix up mere *practical* matters with matters of pure *taste*, as an example of the combined process which should ever be going on in the mind of the architect.

In connection with the subject of drains we should not forget to mention the frequent advisability of constructing a dry drain to preserve the face of the underground walling from damps;* and, as a matter of equal importance, the young architect will not forget the mischief of dry-rot, and the noisome exhalations of foul and stagnant air from beneath such ground-floor or basement rooms as have not a free ventilation beneath their joists, effected by apertures and gratings in the outer walls. [Basements may be effectually ventilated by flues carried up from them at the side of other flues. By having hollow walls (see Part V., *Sanitary Construction*) ventilation can easily be secured by providing apertures under basement ceiling into the hollow. These hollows also keep basement-walls thoroughly dry and supersede the necessity of dry areas.]

We have now disposed of the leading considerations which should be entertained by the young architect in

---

* [A *dry* area formed by a wall slightly battered against the earth, or a semi-arched vault resting against the walls of house, is best in some cases.]

forming his plans and elevations of the general carcass of a dwelling-house, alluding to such paramount practical matters as are necessary to ensure the comfort of its occupiers, without which all merits of architectural propriety and decorative beauty will be regarded as the mere impositions of taste to conceal defective convenience and careless construction. People whose tempers are disturbed by leaks, and offensive smells, by damps and smoky houses, or even by partial failures in design, will become proportionally blind to the numerous merits which may still remain; and the architect, at the moment of signing his last certificate, " that the contractor has fulfilled all his duties in the most complete and workmanlike manner," may be, in effect, signing the declaration of his own inefficiency, and unconsciously entering on a period of much trouble and perplexity, when he imagines the completion of a pleasing labour which is to establish his professional competency, and produce much future employment. To call his attention to the remaining numerous details of his work, the copious stock specification which forms the substantial worth of this volume will, we trust, be found sufficient; or, at least, sufficiently assistant, in conjunction with his own acquired knowledge and sagacity. A studious and repeated perusal of Bartholomew's " Practical Architecture," Part I., will supply all that is here omitted, and will afford not less information on the subject of the Beautiful than on that of the Constructive.

VIEWS OF EMPLOYER.—It will now be as well to afford a few hints as to the relative position of the architect and his employer. The former usually errs in giving to the other a credit for thoroughly understanding his drawings; while the latter equally errs in thinking that his architect is omniscient, not only in the general laws of design, but also in the particular fancies of individual patronage. Now the correction of the employer is out of the question. He must be taken as he is found; and the architect must then find out what he *is*, "hanging clogs on the nimbleness of his own soul, that his patron may go along with him." What the employer *says* is not invariably what he *means;* and he does not always think that he is bound to say *much* to a professor who is supposed to know *everything*. He gives vague ideas of form and size and arrangement which the architect too hastily receives as positive instruction ; and, the result proving wrong, he is abused for not having acted in correction instead of obedience.

Be cautious, then, in the first instance, of receiving as law the dimensions which are given for the required size of rooms. Show your employer an existing room of the form and size he describes, and learn that he means such a room. If you have not assured yourself of this, your troubles will begin before the walls are twelve inches above ground; for he will then declare that the room is much less than the model chosen, and will hardly believe it to be otherwise, in spite of the arithmetic which shall be conclusive that it is so. Nothing is more deceptive than the appearance of comparative smallness in the rooms of a building only plinth high.

OCULAR IMPRESSIONS.—Consider, secondly, that the mode of finishing rooms,—with a heavy, or light, cornice,—a dado, or only a simple skirting,—a plain or a richly decorated ceiling,—materially affects their apparent size; the space taken from the plain part of the walls and ceiling being, in effect, equal to a diminution of actual capacity; or, at least, to an alteration of its proportions. The annexed figures will illustrate this fact. Seen in close proximity, their sizes are observed to be the same; but, looked at in separate succession, fig. 1 will be called long and low, fig. 2 spacious and high.*

Having, therefore, satisfied your employer in *real* dimension, — a question touching only his carpet-room, table, and chairs, —you may disappoint him in the expression of *apparent* dimension, which is a question of optics.

Fig. 1.

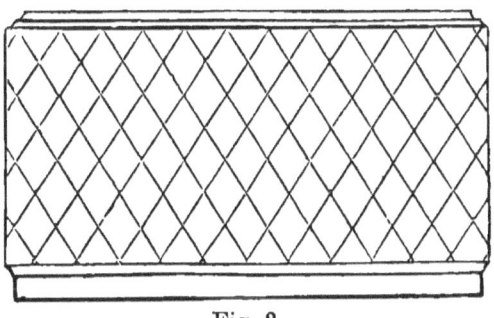

Fig. 2.

Real and Apparent Size.—The *expression* of relative sizes and proportions will, in fact, be the only certain assurance of the patron's complete approval, and thoroughly to effect this, a differing altitude, in rooms, &c., of differing magnitudes in the same house, should be given where readily practicable. If, for instance, corridors of only six feet wide lead to rooms of sixteen feet wide, and both be of the same height, the former will appear too narrow, the latter too low. Both will be advantaged by diminishing as much as may be the height of the corridor; and where the actual ceiling of the latter cannot (on account of the corresponding height of the doors or windows of both)

* See page 193.

be sufficiently lowered, a frieze, or entire entablature, with a higher skirting, or a dado, may so diminish the 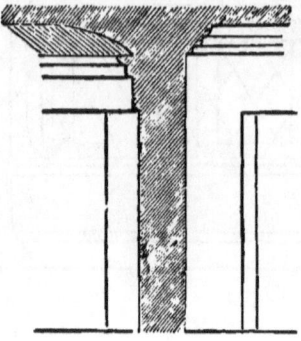 apparent height of the corridor as to give it seeming width, and preserve undiminished the height of the room. Here in fact is the legitimate use of false or pseudo-architecture; and there is no doubt that of two houses, the one well studied in these particulars, and the other not, the former will leave a general impression of spaciousness in all its rooms which in the latter will be wanting.

VALUE OF CORNICES.—Cornices will therefore be thrown flat on the ceiling; brought down upon the 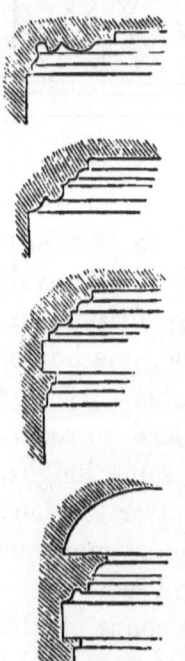 wall; with friezes; or complete entablatures; either immediately connected with the level ceiling, separated from it by a cove, or forming the imposts of a segment or circular ceiling, as the rooms and passages shall be relatively higher or lower. It is not merely to please the eye of the critic that this is done. He will indeed have the double pleasure of seeing it, and of knowing *why* he sees it; but it is sufficient that your employer is well satisfied with the result, without any knowledge of the wherefore.

WINDOWS.—Another point of frequent disappointment to your employer is the width or the height of windows, and more particularly of the position of their sills in relation to the floor.

He may be pleased by their appearance in your elevations, but he may have such particular notions of his own as to their effect inside the rooms, that a material and costly alteration may subsequently increase the bill of extras and affect the beauty of your exterior. Be clear, then, in the beginning, whether sliding sashes or French casements are to be insisted on. Whether the sitting or other room windows are to reach or nearly reach the floor, or whether there are to be backs and elbows; and at what height the glazing is to commence. Forewarn him that French casements are, in this country, scarcely to be rendered weather-tight: that if they open outwards, they are liable to be dashed to pieces by the wind; if inward, they interfere with curtains and defy falling Venetian blinds. Again, if there are to be outer folding shutter blinds, your architraves or other dressings will be in the way; outer shutters, in short, are most hostile to decorative dressings.* In large bay or triple windows, the management of the shutters should be considered even in the first sketch of your plan. They may be too wide to allow of the ordinary folding shutters in side boxings only; there may be objections to form additional boxings against the mullions; lifting shutters, from casings beneath the sill or the room floor, may be required. Leave not these considerations " for the future." Think of them at once. In domestic *Gothic* windows, especially, be careful to anticipate. Mullions and transoms of stone demand the greatest care in forming a weather-tight meeting with casements of metal; and mullions and transoms of wood will be tortured with the alternations of heat and wet. They are fearful contrivances to catch the beat of

---

* Outside sliding shutters, or the kind of blind called "helioscene" are more impermeable to the sun and heat.

weather, and to hold it when caught, till it bubbles through the sash or casement joints, and calls for the housemaid to bring pots and pans to catch the unwelcome stream. Are casements, or sashes, to be used? If the former, the framing will be *solid;* if the latter, hollow *cased.* Are the sashes to rise through the transoms? Are the casements to open inwards or outwards? Nothing may be easier to the architect than such a sketch of a Gothic front as will fascinate his employer's eye.

[The position and design of windows need the most careful thought on the part of an architect. As regards position in a room, they should be designed with especial reference not only to *lighting* but *outlook.** A proportion of 2 to 1 or nearly that height is required for single openings in principal rooms, though a less height may be given in upper stories. The inter-fenestral spaces or piers should as a rule not be less than the width of clear opening, except in cases where wide centre windows are required to give light, or to produce effect. Jambs should be well splayed, and in dark confined neighbourhoods, the heads also should be splayed externally. In bedrooms, windows should be placed as high as possible both for ventilation and appearance; a deep dead space of walling over a window is oppressive and objectionable. As to the filling-in, sliding sashes are preferable to casements in all exposed situations, though they lend themselves with more difficulty to architectural appearance than French or mullioned casements. We have occasionally used horizontally sliding casements instead of hinged ones, and with good effect.]

Sharp and vigorous touches of the pencil, picturesquely showing the moulded recesses of a Tudor window, are readily done; but not so the contrivance

---

* For principal windows, the south-east is perhaps the best aspect.

of such working details as shall in execution preserve the pictorial and keep out the rain. Consider the cost of the long rod-bolts, and other expensive articles of copper and metal, which the pelting of the storm will render imperative. If these "appliances" be not thought of at the first, your employer will surely regard them, in their subsequent adoption, as remedial *extras* to make good radical defect. Our own perplexities in these particulars have been frequent and harassing, and we would put our younger brethren on their guard. Let them recollect, that, in a smart sketch, they draw out and sign a promissory note, which may require all the wealth of their practical attainment to pay.

ARCHITECTS' DUTIES.—A gentleman, in employing an *architect*, will not fail to consider that he *might* do without one; and it is, therefore, under the impression of something over and above what the *builder* could do for him, that he incurs the expense of professional advice and assistance. The least he has a right to expect, is the value of the *artistical*, in addition to that of the *practical;* and he may reasonably expect yet more. Superiority of taste, and of ingenuity in arrangement, he will look for as a matter of course; and it must be admitted that he has also a right to superior knowledge in respect to the economical (and at the same time fully efficient) management of material and general construction. But he will frequently (and not with so much reason) look for more still, and be inclined to visit upon his architect those failures in *particular* construction which only good workmen can insure, under the direction of the contracting builder, and the supervision of an ever-watchful clerk of the works, exclusively occupied on one job. It will be well, therefore, for the architect at once to undeceive his employer in this last particular. He cannot be always present to see that the interior of the walls is well compacted with solid filling and sufficient mortar; that foundations and drains, which are concealed as soon as laid, have been executed in thorough obedience to his specifications; that every slate is properly nailed, and every piece of lead flashing inserted sufficiently in the masonry; that all the carpentry is thoroughly sound and seasoned; that all the joinery is properly "framed, glued, and blocked;" that the plastering has been mixed in the prescribed proportions, and efficiently worked up; that the flues are all of the full size and properly pargetted; that paving has been laid on a well-prepared bottom; in short, he cannot, during only the occasional visits of inspection which he engages to afford, see into those parts of the work which have in the intervals been concealed; nor can he anticipate those future deficiencies, either in work or material, which may not show themselves in any degree until some time after the occupation of the premises. He will have done much in observing, that all, which from time to time remains developed to him, is effected to his satisfaction; and his drawings and specifications will still remain, to justify, under any future chance of impeachment, their sufficiency as a means towards a satisfactory end. Even a contractor, however-practically competent, cannot be always on the spot; and no merely ordinary foreman can be trusted in his stead,—because *if* he be so trustworthy, he is worthy of the double pay which will leave him an "ordinary foreman" no longer.

CLERK OF WORKS.—When an employer will not take upon himself the responsibility of trusting to the efficiency of the contractor and his men, the architect is bound to insist on the engagement of a well-tried clerk of the works. The author of these "Hints" has suffered so much, from a too ready desire to save his employer the charge of a constant supervisor, that he cannot too strongly urge upon those whom he now addresses the advisability of having a clear understanding with their patrons on this point.

DETAILS.—To recur to a few matters of *taste* in the interior finishings of houses. We recommend the young architect to provide a good and varied supply of specimen drawings for the enrichments of cornices, ceilings, internal dressings and panellings of doors and windows; chimney-pieces of marble, for sitting and best bed rooms; of wood, combined with stone or marble, for bedrooms; of wood, with slate slips, to keep the former from the heat of the fire; and of cheap simple chimney-pieces of Portland or slate, for kitchens, offices, and inferior rooms; designs for columned screens, dividing a long sitting-room into two compartments, and supporting a partition wall, or stack of chimneys, in the rooms above; for columned or pilastered decorations for dining-room recesses; for staircase or other lanterns, with their cornices, and the enriched soffits round their openings in the main ceiling; for turned wood, or cast iron, stair balusters; for cast-iron lights over entrance-door transoms.

All these are matters in which the individual fancy of the architect, and the whim of his employer, may be more indulged than in those severer and more conventional features which constitute external architectural decoration and character; and the young professor, during the leisure of his yet only partially occupied time, may advantageously keep up his hand as a draughtsman, and invigorate his imagination as an artist, by studying them, and providing a series of such examples as will, hereafter at least, prove suggestive, if not ready at once for adoption. Employers can rarely see *what* they desire unless they first see something *like* it. The slight sketches of these things which appear in small sections and elevations, or which are vaguely described in specifications, will merely serve as postponements of available consideration, and this will arrive at a period when you may regret not having entertained it before. Against a sketch or slight description, a contractor puts a low and unconsidered price; and when your working drawings are afterwards made out, he considers them as much beyond the thing intended, as the employer thinks them beneath it.

ORNAMENT.—The economy of ornament is not so much shown in employing it only where most needed, as in sparingly employing it, with due relative proportion, in *every* place where it is needed at all. Thus, in all parts of a house which are seen in immediate and unconcealed connection with the principal rooms, their relationship *to* those rooms should be marked. As an instance of prevailing defect in this particular, we may allude to the application of bold and handsome cornices to staircases and the ceilings of staircase landings, while the plastered soffit of the stairs forms a plain and mean-looking junction with the face of the wall. You need not, it is true, continue the modillions of a landing cornice down the rake of the stair soffit to the floor; but you should unquestionably continue down it one or more of the upper mouldings of that cornice.* White plastered soffits are not, in fact, the most suitable to a range of wainscot stairs. Plaster *expresses* stone :† and no one would think of casing the ends, risers, and treads, of a flight of stone steps with

---

* [It is strange that architects overlook the relief of their stair-soffits Why should they not have raking cornices?
† The inference here is hardly logical (see Design).]

wood. The soffit of a wooden flight of stairs should, therefore, either be formed of wood panelling, or of plaster, papered or painted in imitation of it: but, under any circumstances, forget not the raking moulding. Even when the stairs are of stone, with under-cut mould- 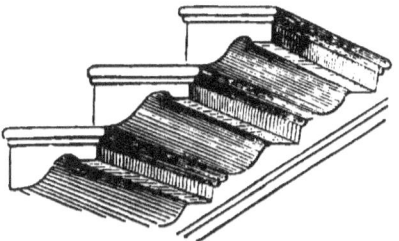 ings, we would still show the moulded work stopping short of the part of the stone inserted in the wall, forming the intervening part into a continuous raking line, and running under it the plaster moulding we have alluded to.*

Hints on Comfort and Convenience.

DOUBLE DOORS.—In allusion to a few matters of *comfort* and *convenience*, we would hint at the virtue of being a match for the occasional violence of gusty weather, in so contriving that two doors shall be passed before you are fairly in the body of the house. Thus an enclosed porch will enable you to shut the outer or porch door, before the inner or passage door is opened. An entrance vestibule should, if possible, have only one outer door, and one inner door leading into the staircase or hall of common internal communication. The interception of through draughts cannot be too attentively considered. A range of doors, all opening one way, in a long passage with a window at each end, will often exhibit the very perfection of the evil; and you may not expect your lady patroness to give much eulogy to the perspective of your corridor if she loses her cap in passing through it, and only gains in return

* [Or the ends of steps near the wall could be left to show their whole section, the soffit being countersunk.]

a sore throat. The air that can quietly and courteously insinuate itself into rooms under the bottoms of the doors, in the chinks between the casements and rebates, and through the fire-place, is a welcome and necessary guest; but when it takes to *slamming* doors, breaking windows, and carrying hearth-rugs up the chimney, it is a symptom of some great want of caution in the architect.

POSITION OF FIRE-PLACE.—The best situation for fire-places in a large room is, unquestionably, in the centre of the longer side; and, for doors, close to the extremity of the same. The worst position for doors is at the extremities of the walls at right angles to the fire-place side, looking directly over the length of the hearth. Where doors communicate between rooms, they will be best placed in that part of the partition walls nearest the window side and furthest from the fire. In smaller rooms, where there is scarcely suf-

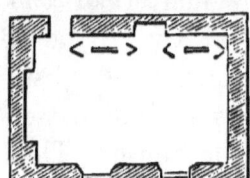

ficient room for a central fire-place, and *two* doors equidistant from it on the same side, it is often better to put the fire opening in the centre of the length between the one door and the end of the wall. It is not only more comfortable, but, where there is no projecting chimney

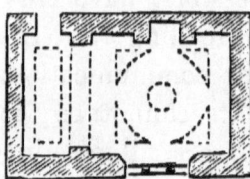

breast, more sightly. Under some circumstances a perfect preservation of centrality may be obtained by the use of breaks or pilasters and ceiling beams, as the adjoined figure exemplifies.

[Doors should be placed in such positions as will least destroy the privacy and comfort of a room. To place a door opposite a fire-place and in the centre of the wall of a room is manifestly bad; it creates a direct draught over the most frequented part of a room, besides the inconvenience of intruding upon those seated round the table or fire. It should therefore be placed either at one end or in one of the corners of the room opposite the fireside.*]

WARMING AND VENTILATION.—If any method of general warming and ventilation be required, it will be for the architect to choose from the number of patents and practices in vogue, and to prepare for them in his first plans. An early conference with the patentees or professors of these methods should be of course secured, and their proposed operations duly provided for.†

KITCHEN AND OFFICES.—The perfection of kitchen and office comfort is, perhaps, a "consummation" more "devoutly to be wished" than any other; for all others are especially dependent on it Expect no master or mistress to be happy, while a cook, housekeeper, and butler are discontented. Keep the smell of the mutton fat, cabbage water and chopped onions out of the main house, for the sake of the hostess and her guests; but, for the united sake of all parties, make your kitchen, scullery, larder, store-room, and pantry replete in all the sufficiency of space, fittings, and communication. Old servants may have accommodated themselves to old defects; but the success of new and better arrangements will, for a length of time, remain problematical. Your only chance is to flatter old servants by consultation. Learn from them the merits and demerits of their present accommodation; submit the result of your ingenuity, and of their exactions, to the upper house; and thence deduce a well-studied plan, to be again modified till you *think* both houses are satisfied. Forget not the cook's closet, the still-room, the china closet. Remember that,

* Take care in planning bedrooms that the door does not occupy the bedstead's place, and that the door be hinged so as to screen the bed.
† See Sanitary Construction, Warming, &c., Part V.

besides a larder for cooked meat, another for hung meat, and a salting-room, may be required. The dairy may be insufficient without a scalding-room and a churning-room. The butler's pantry may be incomplete without a separate glass-washing and plate-cleaning room, and a strong closet for the security of plate not in constant use. Enable the housekeeper to have an eye on the cook, and the means (by a sliding door) of communicating with the kitchen without necessarily going into it. Keep the servants' hall and back entrance out of the way of the operations and runnings to and fro during the bustle of dinner; but, at the same time, "handy" for a speedy advance to the front door of the main house. Remember that a butler's satisfaction is improved by well-arranged cellars for beer, strong ale, wine in casks, and ditto in bins; and that a master's comfort is enhanced by ready access to the said cellars from his own part of the building. Forget not that it will be well if the brewhouse is connected by pipes with the cellars; and especially bear in mind that pipes of wine which are from five to six feet long have to go lengthwise down an inclined plane and through doorways into the cask cellar. A corking-room and a bottle-room follow of course. Let your coal and fuel stores be prompt for the supply of offices and main house, and consider that coals are of at least two qualities, and must exist divided. Let your gallantry think of pretty maids carrying coal-scuttles in their hands without bonnets on their heads, and provide covered ways for their benefit. To the boot and knife house it may be well to add a brushing-room with a good stove in it, or a drying-closet for wet clothes. Let the wash-house yield its cleansed linen readily to the laundry, and the laundry its mangled and ironed ditto to the linen-room. Consult propriety in keeping the maids and the men-servants in a state of respectful separation, with separate staircases to their respective dormitories. Let the housekeeper's and butler's sleeping-rooms respectively command those of the former.

Returning into the main house, we may mention the convenience of a waiting-room connected with the master's private room; and again, connected with the latter, a fire-proof strong closet and a gun closet. A gentleman's bath and water-closet will be well added to this nest of conveniences: a second bath and closet for ladies being provided on the floor next above. On each floor a housemaid's closet will be most welcome, with a pipe from the great reservoir in the roof, to supply each with water for the bedrooms. The convenience of a ready supply of water from one or more reservoirs (to be filled, when not supplied by the rain, with water ejected from the tank below by a force-pump) will be obvious. The water-closets, baths, butler's room, &c., will be jointly dependent on it. The matter of water, though mentioned late in this essay, will be among the first things considered by the young architect, who has, no doubt, an adequate knowledge of well-sinking and steining. It only remains to hint at the policy of providing ordinary closets wherever a recess in the masonry may allow it; for among the stronger impulses of woman is a passion for closets, shelves, rails, and pegs. To crown the ridge of this part of our fabric of hints we simply allude to a good and well-located dinner-bell.

[In large establishments, and especially in towns where several stories are superimposed, the convenience of service lifts is great, and should not be unprovided for. It is easy to provide a corner, a recess, or an irregular space within the kitchen, or serving-room, for such a purpose. Sometimes, if the kitchen is on the ground-floor, a hatchway for serving to the dining-room obviates the necessity of the servants traversing the passages and, perhaps, entrance hall, coming in collision with visitors and guests, and not unfrequently rendering the dinner cold. It is a singular instance of English obstinacy and stubborn adherence to old usage, that the employment of lifts is so uncommon in private houses.]

STABLING, &c.—The usual accompaniment of a good house is a good stable building; and for many useful particulars on this subject, as well as in respect to farm buildings in general, we cannot do better than recommend Loudon's "Encyclopædia of Cottage, Farm, and Villa Architecture."

Let the young architect deny to *none* of his out-buildings the same amount of care which the main structure has had. All should show themselves to be of the same family. The stable court will often afford an opportunity for an advantageous display of simple and expressive architecture. The interior of the quadrangle may exhibit the rough sketch of the more finished external façades, which will admit of much characteristic decoration. For instance, the entrance into the court forming the centre-piece of a range of stables is a most legitimate opportunity for adopting the general *form* of the Roman triumphal arch;* and, in fact, as good and handsome stables are becoming, more and more, one of the signs of squirearchical importance, the young architect will take the subject into his best consideration. The following hints may be of service.

Space enough for a carriage to drive in, turn, and

* However pardonable in its day we cannot approve our author's analogy.

be backed easily into the coach-house, not less than thirty feet wide. Large open porch before stable door for cleaning dirty horses in wet weather. Harness-room between the stable and coach-house, with an opening in the partition wall for an Arnott's stove, common to the coach-house and harness-room. It is most essential that both of these should be dry, and capable of being heated. Boxes not less than the width of two stalls; each stall and box having an efficient ventilating aperture close to the ceiling over the horse's head. In the loft over, avoid flat collar-beam roofs which thrust out the walls. The annexed is the ordinary roof adopted by the author. Let not your smallest stables be less than ten feet high to the under edge of the loft joists. Give larger stables one or two feet more. Avail yourself of the additional air space afforded by the depth 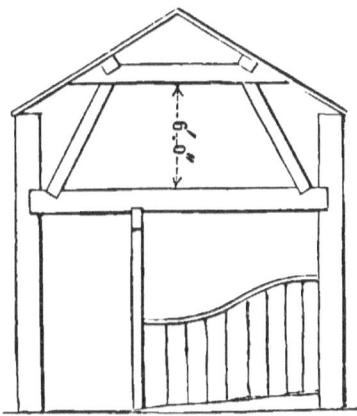 of joists, by having no plastered ceiling; but be careful that your floor boarding above is close, with ploughed and tongued joints. [It is a good plan to ventilate stables by carrying up wooden trunks through roof, with caps.] Let your stall-posts run up to take a head-piece to support the joists of the loft. If one side of your extreme stall be necessarily a *wooden* partition (which is not well), board it to the top: the kicking and movements of a restless horse will shake down plastering. In coach-houses place guard stones, and form tram courses and stops, to prevent the wheels, &c., from rubbing against the walls and one another, and the backs of carriages from being injured against the back of the coach-house. Guard your doorposts with hard stone plinths to direct the wheels on being backed in. Seven feet clear space is ample for a carriage to pass through.

Width of Stables, &c.—Width of stables and depth of coach-houses not less than fifteen, commonly sixteen, and need not exceed eighteen feet. Entrance carriage-way into court not less than ten feet clear.

Stable Fittings.—Gentlemen and their grooms differ in the ideas to which the following queries refer. The young architect will therefore inquire—General dimensions as to widths, depths, and heights? Are the stalls to have surface drainage to trickle into the

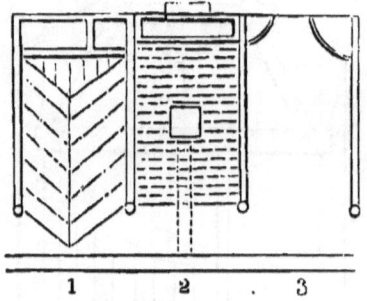

open gutter at foot of stalls, as 1? or a central cesspit and under gutter, as 2? and are they to be flat-paved or pitch-paved? (N.B. Be careful that the open gutter be not a trap to catch the hoof. It should be so wide and shallow as to prevent this danger, or be covered with a grating.) Are the rack and manger to be of wood, level, and near the

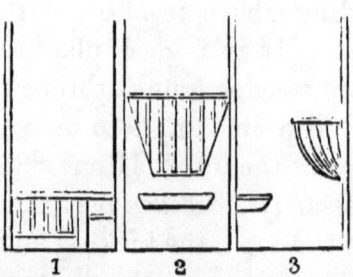

ground, as 1? or one above the other, as 2? or of cast iron, and placed as 3? Is the hay to be thrown down from the loft into the rack, or brought to it from a store on the ground level? Is the corn to be in a chest below? or to be supplied into the stable by a trough connected with the corn store in the loft? (N.B. Case your corn chest with tinned sheet iron, or the rats and mice will work into it.) Is any portion of the loft to be employed for men's sleeping-rooms? If so, there must be regular stairs; otherwise a step-ladder may serve. Consider the position and provision of a door for admitting hay and corn into the loft. Is there to be a crane, or bracket with pulley, &c.? Agree upon the position of the dung-pit, as convenient for the removal of the dung to the gardens or fields. Is it to be so constructed and cemented as to hold liquid manure? Can you unite the sewerage of the house with that of the stables? (Do not conduct the water from the roof pipes and the surface of yard into the liquid-manure tank.) Will the pebble stones of the vicinity afford good paving? or must you procure granite or other spalls for it? Remember that the tread and kicking of horses will soon disturb it, if it be not well bedded, close set, and well rammed. Defects in this particular are of great annoyance. Is a clock-tower required? a pigeon-house? a dog-kennel? Stable privies will, of course, be placed near the dung-pit. Are any open sheds required? carpenter's shop? smithery? well? watering-tank? separate horse infirmary?

[Iron fittings, racks, mangers, open gutters, stall posts, ventilating ramps of cast iron, &c., now generally supersede wooden fittings in stables; they take up less room, are cleaner and less absorbent, and far more durable than wood. One of the best improvements is the adoption of glazed or enamelled iron tiles for lining the heads of stalls and boxes. Iron casements and ventilating openings, and valves underneath the roof or ceiling, should be employed in all cases. For further information the young architect is referred to the various catalogues and works issued by well-known manufacturers of stable fittings.]

For more detailed information on this subject, and

for abundant instruction in regard to farm buildings, &c., we again refer to Mr. Loudon's publications [or Mr. Repton's valuable works].

LODGES : THEIR STYLE.—A word on the subject of *lodges*. They are too frequently more pretending in their architectural appearance than perfect in their internal comfort. A living-room and a little closet of a sleeping-room are crammed into a Portland stone case; and while the Greek order in its full external development emulates an Athenian Propyleum, there is little but *dis*order within from the lack of those absolute conveniences which even the humblest cottager cannot do without. To preserve in cleanly neatness the day-room (in which nobility itself sometimes takes shelter), and to prevent pots and pans from showing in the front of our Doric portico, there must be back rooms and back premises in reasonable sufficiency. The poor lodgekeeper is rarely admitted into those preparatory consultations which are open to the housekeeper, butler, and cook; and it therefore behoves the architect to be his especial advocate. A lodge, in fact, should be a little *house*, and not a little *temple*.* There is not that difference between the lord of the mansion and the keeper of the gate that there is between a Christian and a house-dog. No MAN, who is in the service of his wealthier fellow, should live in a *kennel*, however smart its "complement extern." Occupy, then, that space which will allow of subdivision into room for sitting and sleeping in comfort; for stowage of food, crockery, and fuel; for cooking and cleansing, and other *necessary* matters. Give the lodgekeeper his well and pump, or, at least, his tank and

---

* Since our author's day, a less pretentious and more characteristic treatment is evinced in these buildings, happily for the dignity of Art.

filterer, that he may not be denied the luxury allowed by teetotalism.

In respect to the *style*, we think the *principal* entrance lodge should be a fitting prologue to the "swelling act of the imperial" mansion: true to it, in architectural character, as the mansion portico to the mansion itself. In the secondary and other lodges or cottages, the caprice of a taste for variety may be indulged. *The* lodge at the opening of the grand avenue should, assuredly, be prophetical of the grand structure at the end of it; but there may well be, in other parts of the ground, the *Gothic* lodge, the *Italian* lodge, the *thatched* lodge, and the *Swiss cottage*. The strictest epic poem admits its episode; and the different pictorial aspects of different situations on a gentleman's estate may be suggestive of differing architectural models.

BUILDING COMMITTEES.—In regard to all PUBLIC BUILDINGS, it would be impossible, in a work of this limited extent, to give anything like the detailed instruction which we have ventured to afford in respect to domestic buildings. A few cautionary hints on the manner of dealing with BUILDING COMMITTEES will prove valuable.

We suppose a case.

You have, either by successful competition or immediate commission, obtained the opportunity of executing a church, or any other important structure, in which the opinions of a Board of Directors are to be collected and consulted. The most respectful attention, and the most penetrating efforts to discover the substance of those opinions, is, of course, your duty; and it is not impossible that you may succeed in obtaining some manageable stuff to mould into form; but, under any circumstances, it is advisable you should "oppose your *patience*" to their perplexities, and "arm yourself with a quietness of spirit to meet the probable tyranny of theirs." That they may have chosen *your* plan as the best of many, or that they may have approved your design on general grounds, must not leave you to imagine that the "working drawings" are all you have now to prepare, and the cares of uninterrupted supervision the only ones you have to encounter; since all the raw ideas of improvement, which may successively and suddenly suggest themselves to the various members of the committee, must be disposed of either by the labour of incorporating them with your own, or by long and weighty arguments to prove them "frivolous," if not "vexatious."

NECESSITY OF AUTHORISED INSTRUCTIONS.—Take pains, then, to explain your drawings fully before full committee, and to get the signature of the chairman attached to them before you proceed with the executive. Take especial care that the plans, to which your first estimate had reference, be preserved; and that you obey no injunctions for increasing or altering those plans, without first giving in writing (of which you have a copy) a statement of the addition or deviation which will be thereby occasioned in the cost. Improvements will be constantly desired by unauthorised authority, of which chairmanship takes no current note; and, in the end, memory will only recognise the architect's original estimate.

ALTERATIONS AND ADDITIONS. — Of course your alterations and additions will be shown in a new set of drawings; the original set being put aside with its proper estimate. Be urgent in again and again begging the fullest consideration of your drawings and specifications before you go to public tender, impressing upon the committee the impolicy of making alterations after a tender has been accepted.

DETAILED QUANTITIES.—Make it, however, a condition with persons competing, that the one whose tender is accepted shall give in his detailed quantities and prices. Before it is accepted, require him to abide by any mistakes he may have made in the former, and to allow of any additions or reductions, at their full amount of quantity, and at the prices which he now gives in. It is *his* business to see that the quantities are sufficient; *yours*, that the prices are not too high.

It may be, on the reception of his tender, that there appear reasons for reducing or augmenting the work. If, however, this be not very important in its amount, it may not be necessary to make fresh drawings, &c., as the contract may be concluded in reference to those already prepared, the deviations being shown in making the *working* drawings, being only careful that, before the works commence, the difference of cost be regularly recorded by the treasurer, and that the order for proceeding be signed by the chairman.

DRAWING UP CONTRACT.—We may here remark on the advisability of an architect's being very clear in his notice to contractors, and in his instructions to the lawyer who draws up the contract. As a suggestion for the former we submit the following:—

"To Builders and others. Persons willing to contract for the erection of a ——— at ———, in the parish of ———, county of ———, may inspect the drawings and specifications at ——— ———, from ——— the ———day of ——— until ——— the ———day of ——— now next ensuing. Tenders to be given in not later than ——— o'clock on ——— the ———. The advertisers do not engage to take the lowest tender; nor will any be accepted unless the character, means, and sureties of the person offering it be satisfactory, and the amount of the tender within a certain sum. All further particulars or explanation will be given by the Architect at his office ——— ———."

Dated ———   Signed "

In addition to this, many further cautions and intimations may be necessary, or, at least, advisable; but these may be confined to the architect's office, or the room where the drawings are deposited: and it may be also urged, that all the parties tendering be required to put down their questions on a paper to which all shall have access, so that the replies attached shall by all be seen: the object being, that, in fairness, all may understand the architect's intention alike. State how the legal expenses are to be borne.

FORM OF AGREEMENT.—As lawyers invariably differ in their way of wording an agreement, an architect need be cautious how he ventures to act without their assistance. It is the author's practice to submit to the legal adviser of his employer a printed form, such as experience has enabled him to prepare. It has been often adopted at once, but still he ventures no further than to give its general substance, devoid of legal technicality.

"John Stokes, of ———, in the parish of ———, in the county of ———, having determined to erect a ——— at ———, in the parish of ———, county of ———, according to the drawings and specifications prepared by George Wightwick, Architect; and William Styles, of ———, in the parish of ———, in the county of ———, Builders, being willing to contract for the execution of the works, the said Styles agrees for the sum of £—— to perform them in a complete and workmanlike manner, agreeably, not only with the *letter* of the specification, and in conformity with the drawings *now given*, but also with the full *intention* of the specification, and conformably with other *future drawings* implied by the present, to the satisfaction of the Architect of the said Stokes; it being understood that Stokes will have the right of making any alterations or additions without vitiating the contract, and that the difference in the cost, so occasioned, shall be estimated by his Architect: That, in the event of the Contractor's bankruptcy, or of his failing to proceed satisfactorily, either as it regards time, materials, or workmanship, his employer shall be at liberty to employ other workmen, and to pay them out of the money which may yet remain unpaid to the Contractor, who will acknowledge that the amount of money he may have received before his bankruptcy is to be regarded as full payment for all the work he has done, as well as for all the materials, &c., which may be on the premises at the time; and that if such remaining money and materials are insufficient, the residue must be paid on demand out of the bankrupt's estate: that no money shall be at any time paid, except under the Architect's certificate that the works have been done satisfactorily; and that the Contractor, under certain penalties, shall bind himself to complete the works on or before the —— day of ————, unless the Architect shall justify delay: that at certain stated periods of the work the Contractor shall receive money to the amount of two-thirds (or three-fourths) the value of work done, as the Architect shall estimate it, and that the residue one-third (or one-fourth) shall be paid within ———— months after the completion of the works, provided no defects of workmanship or materials shall have shown themselves in that time; such defects to be remedied by the Contractor before it *be* paid."

GIVING CERTIFICATES.—We need hardly caution the young architect, before he gives a certificate, to be very scrutinising in valuing the executed work *according to the proportional quantity*, or, rather, with reference to the retention of a sum assuredly adequate to complete the building in the event of the contractor's failure. Bodies of men acting for the public have but little compassion for errors of judgment, still less of arithmetic. It is here that the *detailed tender* is of value. Try it, however, before you trust to it, and learn by the trial how to treat it in justice to yourself. Is it hard in an architect to bind a builder to his deficiencies of quantity or price? It is equally hard upon *himself;* for *he receives his percentage only on the amount of the deficient estimate!* It is a point of interest *and honour* in the contractor to take care that he has a proper price: it is a point of *honour only* in the architect—and *against his interest*—to see that the contractor keeps to his price though it be too low.

ARCHITECT'S POSITION.—The position which an architect occupies relatively to his employer and the contractor is often very painful; and the very possible conclusion of a rupture with one, or the dissatisfaction of both, should so constantly be borne in mind, as to prevent too liberal a construction of the contract on the one hand, and too limited an one on the other. There is a natural tendency in all young and ardent minds to trust to the generosity of their patrons, the liberal intentions of their builders, and the favour of circumstances; and it is therefore the more necessary to impress upon the mind of the directing professor, that the patron and the contractor are equally trustful in the acuteness of the architect's foresight and in the clearness of his intentions. It is well to insert in the contract such a " saving clause " as the following :—

"And the Contractor doth hereby admit that the said specifications and drawings are all-sufficient for the substantial and efficient erection of the buildings, &c. : "—

but it will be worse than useless unless the architect, in the event of a dispute, is enabled to show and to prove that they really *are* so. A builder may justly conceive that what *he* understands is what *you* mean; and he therefore unhesitatingly signs the admission you require. Be careful, then, that you do not wrong him by such vague clauses as may, in truth, be little more than concealments of your own practical deficiencies.

With this we conclude our series of miscellaneous hints, and proceed in an attempt to facilitate the operation of the young architect in drawing his attention to such a methodical arrangement of practical detail as may enable him the more assuredly to perfect that most important of all means towards a "binding contract,"—a complete *specification*.

Only let him remember, that we profess to aid his experience, and not to supply his ignorance.

* [It has been ruled that the architect is the agent of the employer for all purposes relating to the building, but it has not been decisively settled whether an architect has authority to order quantities and to bind the employer to pay for them. Architects have by custom an implied authority to do so, but this must always be proved. (See Moon *v.* Guardians of Witney Union.)]

PART IV.

PRINCIPLES OF CONSTRUCTION.

Section I.

THE MECHANICAL PRINCIPLES OF CONSTRUCTION. EQUILIBRIUM OF FORCES.—The following is a summary of the most important principles upon which the balance and stability of structures depend; and which is here given for the convenience of the young architect.

1. *Composition and Resolution of Forces.*—*If two forces acting upon a point be represented in magnitude and direction by two straight lines drawn from the point, then the diagonal of the parallelogram described upon these lines will represent the resultant in magnitude and direction.*

Let the forces act along lines OX, OY (fig. 18), and let them be represented by OA and OB. Complete the parallelogram OACB, then the resultant force will act in the direction of the diagonal OC and be equal to it in magnitude. These three forces are consequently in equilibrium if we suppose the resultant OC to act in an opposite direction in the same line. When we thus find a single force which is equivalent to two others the parallelogram of forces becomes a rule for the *composition* of forces.

Now the resultant OC may be found by trigonometry thus:—

$$OC = \sqrt{OA^2 + OB^2 + 2\,OA \cdot OB \cos AOB}.$$

Its direction may be found by the formulæ,

$$Sin\ AOC = sin\ AOB\ \frac{OB}{OC};\quad sin\ BOC = sin\ AOB\ \frac{OA}{OC}$$

When the two forces act on a point in directions which include a right angle, the case is one deserving

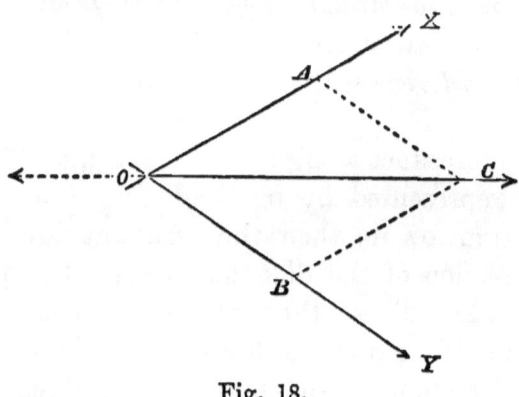

Fig. 18.

particular notice of the architect, as in the thrusts of two arches or frames at right angles to each other. The solution is then simply by Euc. I. 47.

$$R^2 = P^2 + Q^2$$

where R represents resultant, P the force OA, and Q the force OB. For example, if P be 15 lbs. and Q be 8 lbs., then

$$R^2 = (15)^2 + (8)^2 = 225 + 64 = 289,$$

therefore $R = 17$.

The above proposition may be stated in another way, which, from its simplicity and constant use, is of great value in designing structures.

TRIANGLE OF FORCES.

2. *If three forces acting on a point be represented in magnitude and direction by the sides of a triangle taken in order, they will keep the point in equilibrium.*

Let ABC be the triangle, and let P, Q and R be three forces proportional to the sides BC, CA, AB, and respectively parallel to them—P parallel to BC, Q parallel

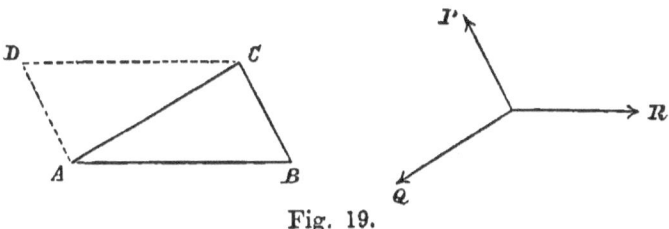

Fig. 19.

to CA, and R parallel to AB. Then forces represented by BC, CA, and AB will keep the point at rest. The dotted lines show how the proposition may be proved by the parallelogram of forces, AB, AD, and CA being three forces in equilibrium.

The above proposition is called "The Triangle of Forces."

The same question may be solved by calculation. Let P, Q and R be the magnitudes of the three forces, and DAB, BAC, CAD the angles between their directions; then,

Sin BAC : *sin* CAD : *sin* DAB : : P : R : Q.

Each of these forces is equal and opposite to the resultant of the other two; or if three forces as above keep a point in equilibrium, each force is proportional to the sine of the angle between the directions of the other two.

Another way of stating the same principle of great use in determining the strains upon frames, as roof trusses and the thrust of arches, is as follows:—

If three forces acting on a point keep it in equilibrium, a triangle having its sides *perpendicular* to the directions of the forces may be drawn, the sides being respectively proportional to the forces perpendicular to them.

Note.—Sometimes cases occur where it is easier to apply the latter method of finding the proportionate strains than that by parallel lines. The principle may be applied in determining the weights for arch stones in equilibrium (see section on Stability).

3. *If any number of forces act at a point and be represented in magnitude and direction by the sides of a polygon taken in order, they will be in equilibrium.*

Let P, Q, R, S be forces acting at a point. From any point A draw AB to represent P in magnitude and direction, draw BC to represent Q, CD to represent R, and DE to represent S. Then EA will represent the

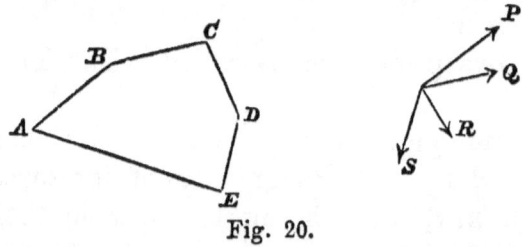

Fig. 20.

resultant in magnitude and direction, and these forces will be in equilibrium. This method is applicable to any number of forces, and is called "The Polygon of Forces." It is often necessary to determine the resultant of a system of forces acting at or through a point, or the thrusts of a polygonal frame of bars (see section on Stability).

Note.—It is not necessary that the forces should lie in one plane.

4. As we can *compound* two forces into one so we

RESOLUTION OF FORCES. 103

can *resolve* one force into two others in given directions as OX, OY (fig. 18). For let OC represent a force. Draw a parallelogram OBCA having OC as a diagonal, then the force represented by OC is equal to two forces represented by OB and OA respectively. Thus, if we assign the directions of the two components we can readily resolve any force into them.

One case deserves special notice.

To resolve a force into two others at right angles to each other.

Let AX, AY (fig. 21) be at right angles, and let a denote the angle RAX, and R the force to be resolved, X and Y the required components, then,

$$AX = AR \cos a \; ; \; AY = AR \sin a \; ;$$
$$X^2 + Y^2 = R^2 \; ; \; (\cos^2 a + \sin^2 a) = R^2$$
since $\sin^2 a + \cos^2 a = 1.$

Thus any force P may be resolved into two others, P $\cos a$ and P $\sin a$, which are rectangular components of the force P.

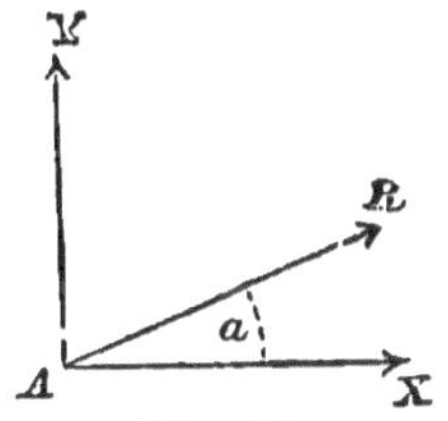

One very general rule of great importance may be stated here.

Fig. 21.

To find the resolved part of a force in any given direction, Multiply the expression for the force by the cosine of the angle between the given direction and that of the force.

Ex.—Let a weight of one ton be drawn up an incline, and let its inclination to the horizon be 4°, then the resolved part of the weight parallel to the incline will be 1 × $\sin 4°$, or 156·25 lbs., which will be the necessary force required, neglecting friction. The resolved part of the weight, supported by the plane, will be

$$1 \text{ ton} \times \cos 4° = 2234·5 \text{ lbs.}$$

5. Another general case may be noticed. To find

the direction and magnitude of the resultant of any number of forces acting in a plane at one point.

Assume any two directions at right angles as axes; resolve each force into two components X, Y, along those axes; take the resultants of the components along the axes separately, and they will be the rectangular components of the resultant R of all the forces, thus—

Let AX, AY denote the rectangular components, then

$$R = \sqrt{AX^2 + AY^2}$$

and if θ be the angle which R makes with X

$$\cos \theta = \frac{AX}{R} \; ; \; \sin \theta = \frac{AY}{R} \; ; \; \therefore \tan \theta = \frac{AY}{AX}$$

When the forces act in different planes, Assume any *three* directions at right angles to each other and resolve the forces along those axes, proceeding as before, or the resultant may be found graphically (3).

6. MOMENTS.—This is a principle of constant occurrence and of great utility in construction, and one which the architect should be perfectly familiar with.

The *moment* of a force with respect to any point, is the product of that force into the perpendicular drawn from its line of action to the point. Thus if AB be the line of action of a force, O any point, and P denote the force, its moment with respect to O is

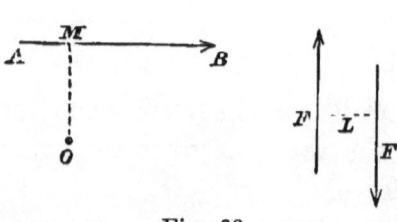

Fig. 22.

$$P \times OM$$

A *couple* consists of two equal and opposite forces acting at right angles to a rigid rod. Thus F F represent a couple of equal parallel and opposite forces; L, the perpendicular distance between them, is called the *arm* or *leverage*, and the product, force × leverage, or F × L is the *moment* of the couple. The tendency of a couple is to cause the plane of the couple to revolve, or to make a rigid rod twist about its middle point.

The following important laws relating to moments should be known :—

Any number of forces acting in the same plane, and any point being taken in that plane, the sum of the moments of the forces tending to turn the plane in one direction about that point is equal to the sum of the moments of those tending to turn it in the opposite direction.

It will be found moreover that if the forces acting upon different points of such a system be transferred to a single point and applied there, parallel to their original directions, they will hold it at rest. Thus forces acting as above are subject to the same conditions as those necessary to the equilibrium of forces at one point; and also that the sums of their *opposite* moments about a given point are equal.

Another interesting property may be noticed : since the area of any triangle is equal to half the product of the base into the altitude, the *moment* of a force may be geometrically represented by *twice* the area of a triangle having for its base the given force, and for its altitude the perpendicular drawn from it to the apex. The following law results :—

If any number of forces acting in one plane, and being in equilibrium, be represented by lines, and their extremities be joined with any point, then the sum of the areas of the triangles formed which have for their bases forces tending to turn the system in one direction shall be equal to the sum of those tending to turn it in the other direction.

The same truth may be stated thus:—The algebraical sum of the moments of two forces round any point in their plane is equal to the moment of their resultant.

Parallel Forces.

7. THE PRINCIPLE OF THE LEVER.—This principle rests upon the preceding, and is one of paramount importance.

Let us take three parallel forces in one plane which balance each other. The simplest illustration of this is a rigid rod suspended from a point C as in the common balance, and having two weights P and Q attached to its extremities A and B. Then if $CA = p$ and $CB = q$, we shall have the proportion—

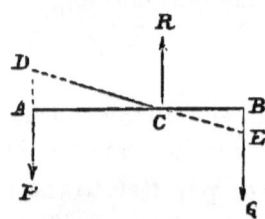

Fig. 23.

$$P : Q :: q : p, \text{ or}$$
$$P \times p = Q \times q;$$

or the relation may be expressed by saying that *the moments of* P *and* Q *about* C *are equal.*

This relation holds when the rod is not horizontal, for by similar triangles

$$CA : CD :: CB : CE, \text{ or}$$
$$P \times CA = Q \times CB.$$

The tension exerted by the third force, or the suspending string (omitting weight of rod), will be equal to P + Q.

Thus to find the relative proportions of three parallel forces in one plane which balance each other we draw a straight line across the lines of action of the forces;

then each force will be proportional to the distance between the lines of action of the other two; or in symbols—

$$AB : BC : CA :: R : P : Q, \text{ and}$$
$$P + Q + R = 0$$

The proposition thus shown is true also of forces not parallel.

Now a lever, which is a rod turning about a fixed point or fulcrum, is acted upon by three forces: viz, the weight to be raised, the power applied, and the reaction of the fulcrum. We may distinguish three kinds of lever:—

1st. When the fulcrum is between the *power* and the *weight*, or a lever of the first kind.

2nd. When the *weight* acts between the fulcrum and *power*, or a lever of the second kind.

3rd. When the *power* acts between the *weight* and the fulcrum, or a lever of the third kind.

In each case we must have moment of power about fulcrum = moment of weight about same point, or in all three cases by the principle of the equality of moments (6).

$$P \times \text{perpendicular from fulcrum} =$$
$$W \times \text{perpendicular from fulcrum}.$$

In levers of the first and second kinds the *resistance* generally exceeds the *power*; in the third it is less than the power.

Since the lever in each case is held at rest by three forces, it follows that the directions of these forces must meet in a point; thus if we produce the directions of the power and weight to meet in a point, the direction of the third force or the reaction of fulcrum is through that point.

Note.—The principle of the lever may be deduced

from the parallelogram of forces, though it may independently be made the basis of statics.

8. CENTRE OF GRAVITY.—The centre of gravity of a body is a point through which the resultant of the weights of its elements passes in every position of the body. Now every body is composed of a number of forces acting in directions parallel to one another but not always in the same plane. To find the magnitude and direction of the resultant of such a system of parallel forces: Find the resultant of any *two* forces in one plane. Then, considering this resultant as a new force, let us find its resultant with a *third* force. This will be the resultant of the first three forces, and may be combined with a fourth, and so on. The amount of the last resultant found is the sum of the component forces.

To determine the position of the centre of gravity of any system of bodies we have simply to assume three different planes at right angles to each other, and take the moments of those bodies from these three planes; the point of intersection of the three resultants thus obtained will be the centre required; for the moment of the resultant of any number of parallel forces, taken from any given plane parallel to them, is equal to the sum of the moments of all those forces from the same plane.

9. *Centres of Gravity of Symmetrical Figures.*—If a plane divides a figure into two symmetrical halves, the centre of gravity will be in that plane; if it can be divided by two planes, the centre is in the line where the two planes cut each other; if the figure can be symmetrically divided by three planes, the centre of gravity will be their point of intersection. A circle, a sphere, an ellipse, ellipsoid, &c., have their centres of gravity at their geometrical centres. These cases suppose the body to be of uniform density and weight.

CENTRE OF GRAVITY.

Compound Figure.—In a figure of two parts, whose centres are known, A and B, draw a line A B. Multiply its length by the magnitude of either of the parts and divide by the whole magnitude; the quotient will be the distance of the centre, C, from the centre of the other part; or,

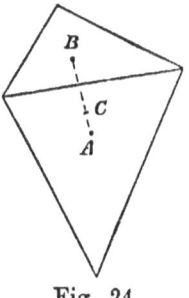

Fig. 24.

$$A C = \frac{B \times A B}{A + B}$$

To find the centre of gravity of two heavy particles or weights: Join the particles by a line A B, and divide it at L, so that A L may be to L B as the weight at B is to that at A.

A Parallelogram.—The centre of a parallelogram is at the intersection of the straight lines which join the middle points of opposite sides, or of the diagonals.

A Triangle.—The centre of gravity of a triangle is found by following rule: Join the vertex or any angular point with the middle point of the opposite side, then the centre is on this line at *one-third* of its length from the side. The centre coincides evidently with the centre of gravity of three equal heavy particles placed at the angles of triangle.

Triangular Pyramid.—The centre of gravity of a pyramid is determined thus: Join any angular point with the centre of gravity of the opposite face, then the centre is on this straight line at *one-fourth* its height from base. The same result holds for any pyramid having a polygonal base, and also for a cone.

[For more extensive information on the subject of Mechanics and the demonstrative part of Statics, the reader is referred to Todhunter's "Mechanics for Beginners," Goodwin's "Statics," Twisden's "Practical Mechanics," and the more exhaustive works of Canon Moseley and Professor Rankine.]

Section II.

10. Balance and Stability of Structures.—No part of the young architect's education, nor the natural gifts with which he may be endowed, which more especially fit him for an artist, as expertness with his pencil and brush, will compensate or make amends for a deficiency in the important subject of the present section, which lies at the very threshold of architectural knowledge, and is the very basis upon which fine-art architecture rests.

A thorough acquaintance with the preceding section, and the application of those principles to actual structures in brick, stone, timber, iron, &c., is necessary before any attempt is made to design structures upon whose safety depend the lives of many engaged in their erection, as well as of those who occupy them.

For a complete knowledge of construction, the reader is referred to Professor Rankine's works, or he may consult with advantage the various works on construction in Weale's Series, and afterwards make himself familiar with the able treatises of Tredgold, Barlow, Hodgkinson, and others, on their special branches.

It is simply proposed here to offer a few hints and simple methods of testing the stability of structures, as abutments, walls, piers, arches, and the means of finding the strains on timber and iron constructions in a few ordinary cases. Practical architects, it is true, seldom resort to such methods of testing their work, and others do not take that trouble: in the former case actual experience and precedent is relied upon; in the latter the indifference proceeds from ignorance, or what is often the case, from a repugnance to what is deemed irksome and laborious. The young architect cannot be too forewarned against this snare. It proceeds from an ardent enthusiasm for the more artistic part of his profession, which often despises the critical, mathematical, and precise for the ideal result; it allows the merely emotional to dominate, frequently to the neglect of truth and science. Let him, however, never neglect the essential, the mathematical, and exact for the sensuous impression, which is often unreliable; the trouble the examination gives tends to verify and correct the æsthetic ideas, and may save the young architect the remorse of an unexpected failure or the ruin of his reputation. These remarks are more called for now, when the young enthusiast is so apt to consult his *eye* and *imagination* solely, and to place an unreserved faith in them.

11. STABILITY OF WALLS, &c.—The application of the Principle of Moments (Section I.) to the equilibrium and stability of walls is one of great importance.

Let A B C D (fig. 25) be a section of a wall, acted on by a

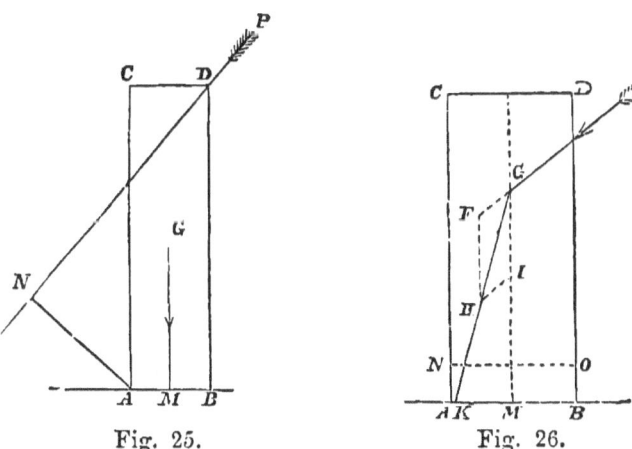

Fig. 25.   Fig. 26.

pressure P, in the direction P N; it is clear that if we neglect the adhesion of the mortar at any bed-joint as A B, and consider the wall as a continuous mass, there will be a certain pressure of P, which will just tend to turn it round the point A, and this pressure will be the greatest the wall can support. This pressure is determined by the rule that its moment about the point A shall equal the moment of the weight of the wall about the same point.

Ex.—Let a wall of brickwork (fig. 25), 18 inches thick and 25 feet high, sustain on the inner edge of its summit a pressure on each foot of its length at an angle of $60°$ with the horizon. To find the greatest pressure that the wall can bear without being overturned proceed as follows:—Draw the section of wall to scale. Make the angle B D N equal $30°$ which the direction of pressure makes to the wall, draw A N perpendicular to P N, and through the centre of gravity of wall G draw the vertical G M, cutting its base in M, then by the Principle of Moments, $P \times A N = W \times A M$; W being the weight, which if we put at 4200 lbs., the pressure P will equal by construction 286 lbs., or that force which just balances the weight of wall. From such a determination it is easy to increase the thickness of wall at its base, especially to ensure the requisite amount of stability.

BUTTRESSES.—It will be observed that by placing buttresses at intervals along the wall the effect is to diminish the moment of P, and to increase the weight of wall; also if the weight of the buttresses is considered, the moment of the weight of wall is increased. In most buildings the thrust at the summit of wall is concentrated at the buttresses, as in the case of groined vaults, and the intervening walling bears little or no pressure; in such cases, if in equilibrio, the moment of lateral thrust equals that of the weight of buttresses. The plates upon which the rafters of a roof pitch, however, tend to equalise the thrusts, and each buttress with its adjacent halves of walling must be taken into the calculation in finding the moment of weight.

The pressures to which walls are generally exposed

STABILITY OF WALLS. 113

are those occasioned by the thrust of arches, vaults, and roof principals.

MODES OF FAILURE.—Walls may yield in different ways; 1st, by one portion sliding upon the other; 2nd, by some portion turning over about one of the edges or bed-joints; 3rd, by crushing of the material, the pressure exceeding its cohesive strength. In the method above given, the second of these failures only is considered, the resistance offered by the friction of the mortar being generally sufficient to prevent the mere sliding of the material; therefore, in taking the moments, the lowest bed-joint should be taken, or that just above the ground. Walls often fail from the ground yielding under the pressure caused by a thrust; the wall then becomes a powerful lever.

Another way of regarding the stability of walls is as follows:—

Let A B C D (fig. 26) be a wall sustaining a pressure P, acting in the direction P G, and let G M be the direction of the weight of wall, acting through its centre of gravity, intersecting the former pressure in G. Set off G I to represent the weight of wall, and G F the pressure; then the diagonal G H will be their resultant, acting in the line G H K.

Let N O be a joint; then if the angle K G M, which the resultant makes with the perpendicular, be greater than the limiting angle of resistance,* the upper portion will slide upon the lower, N O A B (friction being neglected); if, on the contrary, the angle made by the resultant is less than the limiting angle of resistance, the parts cannot slide upon each other, and

* The "limiting angle of resistance" is the greatest angle which the reaction of a surface in contact makes with the normal to it, without motion. Its value varies for different materials. It is of great importance in the investigation of stability of arches and piers (see Friction).

the stability will be greatest if the resultant, G K, is perpendicular to this joint, or ensured if it passes within the base of wall. If it passes *without*, the wall will be weakest at the nearest joint, as N O, the upper part tending to turn on its outer edge N.

It is evident the pressure upon such a joint, when the resultant acts near one edge of the wall, is one of varying intensity; and in designing abutments and lofty towers, or chimney shafts, the resistance offered by the material at such an edge must be sufficient to prevent its crushing; hence the stability of walls and piers, &c., depends upon the magnitude and direction of the resultant of the pressure from without, as thrust, force of wind, &c., and the weight of structure itself combined. Inattention to this, or the supposition that the whole base or section of a wall receives the pressures, has been the source of many lamentable failures, and accidents to life and property.

Professor Rankine gives the following rules:—

Given the load on a bed-joint and the position of the centre of pressure, to find approximately the intensity of pressure at the nearest edge.

In abutments of arches, divide *twice* the load by the area of the bed; in retaining-walls multiply the breadth of the bed by *once and a half* the distance of the centre of pressure from the nearest edge, and with the product divide the load; the quotient is the required intensity.

The intensity thus found should not exceed *one-eighth* of the crushing strength of material. The moment of stability of an abutment may be calculated thus: Multiply the superincumbent weight of mass above bed-joint by the distance of the vertical line through centre of gravity, from the limiting position* of the centre of pressure of the bed-joint.

* *i.e.* safety against overturning.

Practical experience has fixed the greatest deviation of the centre of pressure from the centre of wall or pier at from 0·3 to 0·375 the whole thickness of wall at the bed-joint.

THE LINE OF RESISTANCE.—It is obvious that the stability of a wall or pier, under any pressure, depends on two conditions being fulfilled—1st, that its surfaces of contact, or joints, should not slide upon one another; and 2nd, that the pressure should not be such as to cause any portion to turn upon its edge.

Let A B C D represent any structure with horizontal courses, 1 2, 3 4, 5 6, &c., and let it be subjected to any pressure P, along the line P a. Now this pressure at every joint is compounded with the weight of the stone, or mass, above it. Thus the pressure P a produced would cut the joint 1 2, in a', but the weight of mass A B 1 2 is another pressure, and the resultant (R) of these two pressures would evidently cut the joint 1 2 at some determinate point as b. Now the direction of $a\,b$ is b', but the total pressure will be as before, the resultant (R_2) of R_1 and the weight of mass 1 2 3 4, and so on, the original force, P, being deflected from the straight line at every joint, forming a polygonal line, $a, b, c, d, e,$ as the resultant of the insistent pressures. This line has been called by Professor Moseley the "line of resistance," the resisting points of the different surfaces being all in that line.

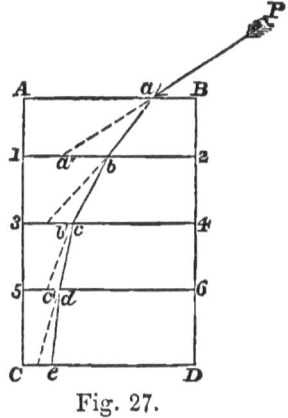

Fig. 27.

Now it is a condition of stability that this line of resistance shall intersect every joint of the structure and pass *within* the mass of it. The line of resistance continually approaches the extrados or outer surface of a wall; hence, if its lowest joint is cut by this line the wall is stable; if it cuts the extrados, unstable. The line may be shown to be a hyperbola, though its determination in walls of ordinary section is seldom required.

12. PRESSURE OF WIND.—The usual allowance for the force of wind in Great Britain is 55 lbs. per square foot. To find the pressure against a circular chimney or tower, take half the area of its vertical cross-section and multiply by the above pressure per square foot of area; the resultant of this pressure will act at the centre of magnitude of the vertical area exposed. The moment of the pressure may be found by multiplying its amount by the height of its resultant above base.

13. PRESSURE OF WATER.—The pressure of water against each foot in breadth of a vertical plane is found to
$$= \frac{\text{depth}^2}{2} \times \text{the weight of water (62·5 lbs. per cubic foot)}.$$

14. RETAINING WALLS.—The *centre of pressure* of a rectangular vertical wall, subjected to the pressure of earth or water, is at *two-thirds* of the total depth from the upper surface, and the direction of the pressure is horizontal.

To find the thickness for a vertical retaining wall of rectangular section supporting a bank of earth of the same height,

Let w' be the weight of earth; ϕ its angle of repose or natural slope; $\frac{p'}{p}$ the ratio of lateral to vertical pressure of the layers of earth; $h =$ height of wall, $w =$ its weight; and $q =$ ratio of intended deviation of centre of pressure from centre of base to the required thickness, t. For bank in horizontal layers—

Functions of ϕ.

$$\frac{p'}{p} = \frac{1 - \sin \phi}{1 + \sin \phi} = \begin{array}{ccc} 25° & 30° & 35° \\ ·406 & ·333 & ·271 \end{array}$$

$$\frac{t}{h} = \sqrt{\left(\frac{w'p'}{6qwp}\right)}; \text{ if } q = \frac{3}{8}, \text{ then } \frac{t}{h} = \frac{2}{3} \cdot \sqrt{\left(\frac{w'p'}{wp}\right)}$$

In a reservoir wall $\dfrac{p'}{p} = 1.$*

The masonry or brickwork of retaining walls should be diminished by steps, the diminution being filled with solid material of equal weight. The face of the wall should be battered (straight or curved).

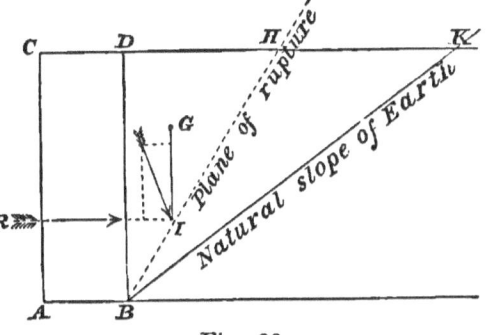

Fig. 28.

TABLE OF NATURAL SLOPES OF EARTH.

	Angle of repose ϕ.
Dry sand, clay, and mixed earth	37° to 21°
Damp clay	45°
Wet clay	15°
Shingle and gravel	48° to 35°

The most usual slopes are 1½ to 1, and 2 to 1.

PRESSURE ON FOUNDATIONS.

	Tons on square foot.
Rock, moderately hard	9·0
Good concrete	3·0
Soft rock	1·8
Firm earth, clay, clean dry gravel, sharp sand prevented from spreading.	1 to 1·5

---

* Fig. 28 shows a retaining wall supporting a bank of earth just on the point of slipping; H B is the plane at which the earth D H B separates. The pressures acting upon this wedge of earth are its weight denoted by G I, and the resistance of wall R, the resultant of these being the pressure against the plane H B, this resultant making with the perpendicular to H B, the limiting angle of resistance. The greatest pressure is found to be when the angle D B H = ½ D B K; hence it is, as the weight of earth is to its pressure against the wall, so is the height of wall to D H; and the pressure against wall $= \dfrac{D H^2}{2} \times$ weight of cubic foot of earth acting at a point *one-third* from base. The formula usually employed is, Horizontal resistance $= \dfrac{w h^2}{2} tan^2 \frac{1}{2} a$; w being weight per cubic foot of bank; $h =$ height of wall, and $a =$ angle of repose with vertical.

Dams, Dock and Quay Walls.—The pressure of water increases with its depth simply; therefore, to determine the pressure against any surface, whatever its position, vertical, inclined, or horizontal, multiply the area of surface by the depth, in feet, of its centre of gravity, below surface of water, and by 62·5 lbs., the weight of a cubic foot. The pressure at any point = *depth in feet* × $62\frac{1}{2}$.

15. Arches of Brick and Stone.—Few structures have received so great an amount of attention and learned theory from mathematicians as the arch. Various theories have been propounded, though generally the practical conditions have been neglected; the arch stones, or "voussoirs," being regarded as smooth bodies devoid of cement and free from friction. The elaborate theories of such writers, while interesting, as evincing much ingenuity and mathematical investigation, and as a study of a speculative kind, must be regarded by the young architect as useful only in directing his mind to conceive the mechanical conditions involved, and to enable him to distribute his weights, and to proportion his structure to the greatest advantage.

Arch of Equilibrium.—The arch of equilibrium is thus far a useful problem. Perhaps the best theory on the subject is that of Canon Moseley, published in his able work on "The Mechanical Principles of Engineering and Architecture," to which the young architect is referred for a complete investigation of the subject of the arch.

In actual practice the element of the cohesion of the cement must be taken into account, and the mode in which an arch actually fails gives us the best means of determining the stability required.

When an arch is not in a condition of equilibrium, or overloaded, and the action of the cement is considered, it is found to yield by turning upon certain joints, and separating into four segments, or portions, as in fig. 29. This mode of failure is caused by the *line of pressure* * cutting the extrados at the crown A, and the *intrados* at the points B and C, called the " points of rupture," the arch separating at the nearest or weakest joints to such places,

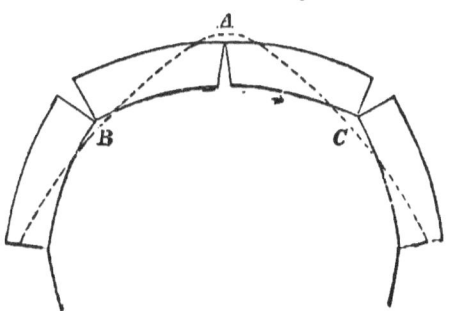

Fig. 29.

some portions turning upon their inner edges, B and C, as on a pivot, and some upon their outer edges, as at A, thus causing the arch to sink at the crown and rise at the haunches. Experiments have shown this to be precisely the case. It is evident that the line of pressure is dependent upon the weight or amount of the horizontal force at the crown. The greatest value of this force consistent with equilibrium is that which causes the line of pressure to touch the extrados; its least value when it touches the intrados, and when the pressure upon the keystone is the least possible which would support either semi-arch. It is evident also that the horizontal force at the crown is increased by the load over it, and that the stability of the arch is dependent upon the depth of keystone and he cohesive resistance of the material.

Again, if there be a deficiency of weight at the crown, the line of pressure falls below the intrados at the crown and rises above the extrados at B and C, causing a depression at the haunches and a rising at

---

* The *line of pressure* is the direction of the resultant of the pressures on the different joints of arch stones (see equilibrium of polygonal frame of bars).

the crown. It is thus seen that there are certain limits within which the stability of an arch is secure, as when the line of pressure is *within* the substance of the arch ring, or does not pass without its boundaries; but that when this line deviates so far as to cut the extrados and thus destroy the equilibrium, the stability is endangered. The greatest stability is ensured, therefore, by the line of pressure being made to cut all the joints, or to pass within the middle portion of the arch ring. Professor Rankine observes it should be within the *middle third* of the arch thickness.*

Hence the stability of an arch is *directly proportional to the depth of its keystone multiplied by the crushing strength of the material*, and is inversely proportional to its radius of curvature multiplied by the weight on each square foot of the keystone

Professor Rankine gives a rough approximation to the horizontal thrust of an arch as follows:—

Take the weight of the vertical load supported between the crown of arch and that point in the arch ring where its inclination to the vertical is 45°.

The stability of an arch may be tested as follows:—

Let A B C be a semi-arch, D E F its abutment, and C the joint of rupture. Through the centre of gravity of the load between the crown A and the point of rupture, C, draw a vertical line. Then, if from a point as M in that line, two lines can be drawn, one as a tangent to the

---

* The line of pressure in an arch is generally that of an equilibrated or inverted catenary, or "curve of equilibrium," and under vertical loads it is very suitable. If a linear arch or rib balanced under the real forces which act on the real arch can be drawn within the middle portion of the arch ring, the stability of the arch is secure. The linear rib or ideal curve so drawn may be the hydrostatic arch or that suited for equal pressures. It is necessary to assume such linear arch parallel to the intrados of the proposed arch. For methods of describing these the reader is referred to Rankine's "Applied Mechanics." Such precautions are, however, unnecessary in designing arches of ordinary span, in which the bond of brickwork and mortar are taken into account. A pointed arch requires a greater load at the crown than a circular or flat arch, but less abutment.

THEORY OF THE ARCH. 121

assumed arch curve or centre line at the joint of rupture, and the other horizontal, so that the former line shall cut the joint of rupture, and the latter the vertical joint at the crown in points which are both within the *middle third* of the arch ring, the stability of the arch is secure. In this case the stability is tested to the point of rupture only; below that point it is assumed to be stable.

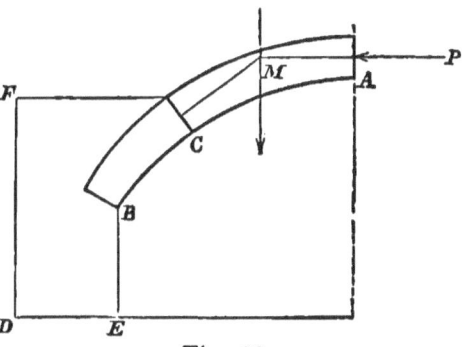

Fig. 30.

For the stability of the abutments, the rule for finding that for piers (Sect. II. 11) is applicable; the horizontal pressure of the arch being first found, the stability of the abutment will be measured by the moment of its weight about its outer edge at the base; or, to ensure stability, the moment of its weight about the said point added to that of upper segment, should *exceed* the moment of the thrust or weight of the opposite semi-arch, taken acting at the point of rupture; or the sum of the moments of the weight of pier and surcharge should *exceed* moment of thrust.

Practically, and regarding the bond and cohesion of mortar, the active thrust of an arch is as the horizontal component, A B, of the load on an inclined rigid bar, A C, fig. 31, from the point of rupture to the extrados at the crown, or as the thrusts of two inclined beams.

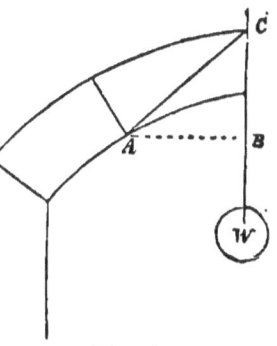

Fig. 31.

The chief object in investigating the stability of the arch is to find the requisite thickness of the abutments to resist the thrust, it being premised that the pressure upon any joint between the voussoirs should be such as not to exceed a safe resisting power of the materials of the voussoirs.

For an investigation into the subject of arches and domes the reader is also referred to the concise work of Mr. Tarn on the "Science of Building."*

Angle of Rupture.—In circular arches the angle of rupture is found to lie between 45° and 55°; and hence, if the squared backing is carried up to an angle of 45° to the horizon it will be sufficient.

Depth of Keystone.—Rankine gives from 20 to 40 as good medium values of the excess which should exist beyond the depth of keystone necessary to resist the crushing of the material, as founded on the best existing examples. He also gives the following empirical rule:—

Take a mean proportional between the radius of curvature of the intrados at crown, and a constant whose values are as under for the depth of keystone, or Depth of keystone for a *single arch* in feet

$$= \sqrt{(\cdot 12 \times \text{radius at crown})}.$$

Depth for an arch of a series as an arcade,

$$= \sqrt{(\cdot 17 \times \text{radius at crown})},$$

or "for equally intense external loads and equal angles of rupture, the square of the thickness of keystone should vary as the radius of intrados."

ABUTMENTS.—In some of the best examples of bridges the thickness of the abutments ranges from 1-3rd to 1-5th of the radius of curvature of arch at its crown.

PIERS.—For piers of a series of arches, the common thickness is from 1-6th to 1-7th of span. All

* Lockwood & Co.

large abutments should be built hollow, with "jack" arches, and have inverted arches at the foundations to distribute the pressure.

16. CUPOLAS OR DOMES depend upon similar principles for their stability as the arch, as a dome may be considered to be composed of a number of portions, or vertical ribs, cut by planes whose intersections are in the vertical axis of the dome. All domes exercise a horizontal thrust upon their supporting walls or "drum," and therefore require iron bands, or chains, inserted at, or just above, the springing. The weakest joint of a hemispherical dome is found to be that which makes 20° with the horizontal; and it is at this joint the thrust is greatest.

In Gothic or Pointed domes, the angle is smaller.

Stability and Strength of Beams.

17. CASE I.—Suppose a beam (fig. 32) supported at its ends, A and B, and loaded at some intermediate point, L. Let the supporting pressures and direction of the load be parallel to each other, as in figure, where P represents the resultant of the load, including weight of beam. Then, according to the principle of the lever (Sect. I. fig. 23), each of the three forces is proportional to the distance between the lines of action of the other two, the load being the sum of the two supporting pressures, or—

$$P : R_1 : R_2 :: \overline{AB} : \overline{LB} : \overline{LA}, \text{ and}—$$
$$P = R_1 + R_2$$

CASE II.—Let the beam be inclined (fig. 33) so that the pressures and load are not parallel as A B. As the forces are in equilibrium, their lines of action must meet in one point, o, and are proportional to the three sides of a triangle respectively parallel to their directions.

To find the directions and magnitudes of the supporting forces. — Let P be the resultant of load or weight of beam acting through its centre of gravity, L, and A and B the supporting forces. Produce the line of action of B or R_1 till it cuts the vertical line OLP in O. Join OA and it will be the line of action of the support A.

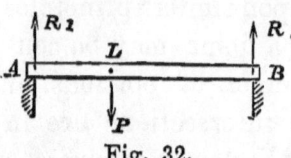

Fig. 32.

To find the forces.—Set off on the vertical line OP the weight as OL to scale, and through L draw a line parallel to the horizontal pressure, R_1, meeting OA in R_2, then the sides of the triangle thus formed will give the ratios of the forces thus—

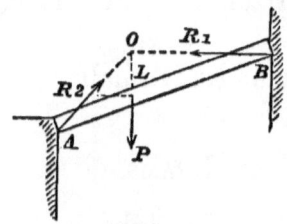

Fig. 33.

OL = weight; OR_2 = supporting pressure, A; and LR_2 = supporting pressure, B. Or, algebraically, if a and b denote the angles made by the lines of action of the supports with the vertical line of load, then—

$$P : R_1 : R_2 :: \sin(a+b) : \sin b : \sin a.$$

Case III.—One of the most useful cases is that of two beams abutting against each other at their upper ends, and against two walls or supports at their lower ends, and loaded at the apex, as AB, AC, fig. 34.

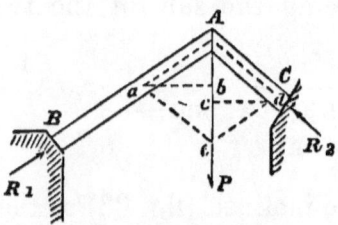

Fig. 34.

To find the strains along each beam and the thrust against walls.—By the parallelogram of forces (Sect. I.) resolve the load along the beams thus: make Ae equal the load; draw ea parallel to AC and ed parallel to AB; also draw ab, cd perpendicular to Ae; then the strain on AB $=$ aA or ed, and that on AC $=$ Ad or ae. Each of these strains can be resolved also into vertical and horizontal components thus:—Aa into Ab acting vertically, and into ab acting horizontally; also Ad can be resolved into Ac acting vertically, and cd horizontally; or, by symbols, putting a and b for the respective angles made by the beams with the vertical or load—

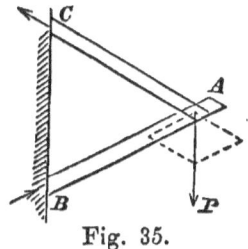

Fig. 35.

$$P : R_1 : R_2 :: \sin(a+b) : \sin b : \sin a.$$

The amount of weight on each wall depends on the inclination of the two beams; in the figure the wall at C sustains the greatest weight; the horizontal strains, however, will be found equal.

18. EFFECT OF POSITION.—In designing any assemblage of beams or bars, the relative positions of the pieces must be considered. For example, in the case of a crane, we have a strut and tie, one piece sustaining a compressive and the other a tensile strain; and these will depend on their relative inclination to the action of the load.

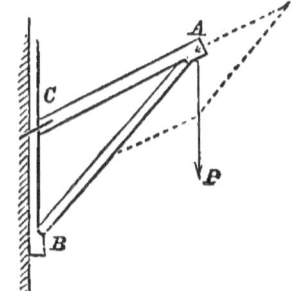

Fig. 36.

Thus in fig. 35, AB sustains compression, and AC tension, but both strains are equal.

. In fig. 36, the strains are greatest with the same weight, and the piece BA has the greatest strain, as will be seen by the dotted parallelogram.

The young architect cannot be too particular in

considering the relative strains which different positions create, by constructing the parallelogram of strains for himself, and thus determining the kind of joints or abutment necessary, and the sectional areas of his timbers to support these strains.

It will be readily seen that the last case noticed, that of a crane, is of common occurrence in roof-framing, as in the instance of a principal rafter and strut, or that of a hammer-beam and strut. It is particularly necessary to discriminate carefully between pieces subject to compression, tension, or cross strain.

NOTE.—It will be observed that there are two methods available for finding the strains in structures, the first and simplest being that by constructing the parallelogram or triangle of forces called the "graphical" method; the second and most accurate by the application of trigonometry, or by calculation (see Sect. I.). For all practical purposes the graphical method is sufficiently accurate, is less liable to error, and one preferable for the architect from the facility it can be used upon the working drawings themselves. In the other method tables of sines and tangents are required.

FRAMES AND TRUSSES.

19. A frame is composed of bars or beams connected together at their ends, the beams being thus fixed in their relative positions. It is clear that the joints themselves offer little aid to the rigidity of such a frame, so that we may consider the pieces capable of turning about them unless fixed at both ends. The points about which such movement would take place are called the *centres of resistance* of the joints, the straight lines joining these centres being the *lines of resistance* of the beams, along which the thrusts or tensions are

TRIANGULAR TRUSS. 127

propagated. If subject to a thrust the beam or bar is a *strut*, if to a pull a *tie*. In the design and construction of frames or trusses it is essential to avoid cross strains, and so to dispose the pieces composing a frame that they shall be struts and ties only; this is the science of good framing, as it is evident, by confining each piece to compressive and tensile strains, the material is most economically employed, the sections or scantlings of the pieces being designed to meet such strains only.

TRIANGULAR FRAME.—The ordinary form of a frame is that in which the pieces form a triangle or a system of triangles.*

Let A B C (fig. 37) represent a frame of three bars or pieces connected at the angles. Let a load P be applied at the apex, and the supporting forces Q and R be those of the walls or supports, then their lines of action must either be parallel or intersect in one point. It is evident these three forces balance each other, and are therefore proportional to the three sides of a triangle respectively parallel to their directions.

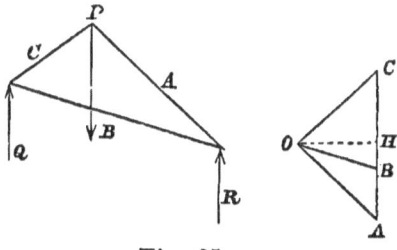

Fig. 37.

When the three forces are parallel to each other, as in the case of an ordinary roof frame, the triangle of forces becomes a straight line, CA.

Draw O A parallel to A, O B to B, and O C to C, then if the load CA be applied at the apex of frame, A B applied at R, and B C applied at Q, are the supporting forces, and the lines O A, O B, O C represent the strains on the bars A, B, C respectively.

* The triangle is the simplest and only figure, the form of which is perfectly rigid, its sides remaining unaltered.

From o draw OH perpendicular to CA, then that line represents a component of the strain which is equal in each piece, and is the *horizontal thrust* of the frame.

The trigonometrical relations are as follows:—

Let a, b, c denote the respective angles of the bars A, B, C to the horizontal line OH, then—

$$\text{Horizontal stress OH} = \frac{\text{load CA}}{\tan c \pm \tan a} \quad (1)$$

$$\text{Supporting forces } \begin{matrix} \text{AB} = \text{OH} (\tan a - \tan b) \\ \text{BC} = \text{OH} (\tan b + \tan c) \end{matrix} \quad (2)$$

$$\text{Strains are} \begin{cases} \text{OA} = \text{OH} . \sec a \\ \text{OB} = \text{OH} . \sec b \\ \text{OC} = \text{OH} . \sec c \end{cases} \quad (3)$$

In a polygonal frame a similar method is to be used, the strains on the several pieces or bars being found by drawing radiating lines parallel to the bars from a point to the vertical straight line representing the parallel forces or the load and supporting reactions; the segments into which these radiating lines divide the vertical line will then represent the forces which respectively act at the joints of every two consecutive bars, and the whole vertical line represents the total load.

The relations of the forces are as follows:—

Let the horizontal component be denoted by H, and let a and b denote the angles of inclination of *any two* bars to the horizontal; and R, R_1 the respective strains along the two bars; and P the resultant of the external forces acting through the joint between the two bars, then—

$$P = H (\tan a + \tan b)$$
$$R = H . \sec a; \text{ and } R_1 = H . \sec b.$$

20. ROOF TRUSSES.—It is often required to determine

the strains which the several pieces of a roof truss are subject to, for the purpose of ascertaining the requisite scantlings of the timbers · or, in the case of an iron roof, the sectional areas of the rafters, struts, and ties.

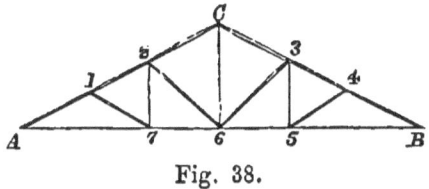

Fig. 38.

Let figure 38 represent an ordinary king and queen rod truss either of iron or timber.

The strains on the main triangle can be found by diagram as before, making the vertical line represent half the gross weight on the roof (see Art. 19); or, algebraically, let w be gross weight of truss, together with the division of roof which it carries, or from middle to middle of adjacent bays :—Let c = half span of truss; k = its rise; i = inclination of rafters; H = tension along tie beam; T = thrust along each rafter; then

$$\text{H} = \frac{\text{w}}{4 \tan i}, \text{ or, } \text{H} = \frac{\text{w}c}{4k}; \text{ T} = \sqrt{\text{H}^2 + \frac{\text{w}^2}{16}} = \sqrt{\text{H}^2 + \text{H}^2 \frac{k^2}{c^2}} = \text{H} \sqrt{1 + \tan^2 i} = \text{H} \sec i.$$

These give the strains on the *primary truss*. For the *secondary trussing* we must find the load on the points 2 and 3; the secondary trusses being A 2 6 and B 3 6.

DISTRIBUTION OF LOAD.—The weight of roof is distributed over the points A, 1, 2, C, 3, 4, B; *one-twelfth* resting on each point of support A and B, and *one-sixth* on the five points 1, 2, C, 3, 4. The loads at 2 and 3 are each $\frac{1}{4}$ w, or half of load between 2 and C, and half of load between 2 and A, or *one-half* between A and C. The strains to be found by equations (1) (2) (3) of article 19. The *smaller* secondary trusses are A 17 and B 45, and each of the points 1 and 4 sustains $\frac{1}{6}$th w.

Tensions.—The main suspension rod c 6 sustains $\frac{1}{9}$ w, and 1-12th w hangs by the queen rods 2, 7, and 3, 5.

The tension between 7 and 5 of tie beam is the sum of the tensions due to the primary and secondary trusses; the tensions at the ends between A and 7 and B and 5, are those due to all three trusses.

Thrusts.—Again, the thrust on c 2 is due to large truss; that on 2, 1 to the primary and secondary trusses; and that on 1 A to all three trusses.

In trussing roofs, &c., the following maxims should be observed:—

1.—The main truss should be composed of one or more large triangles, the simpler the arrangement the better.

2.—Avoid multiplicity of joints, and give to each piece its proper joints.

3.—Design the frame so that every load or strain should be met by a *direct* support or line of resistance, either compressive or tensile.

4.—Avoid cross strains.

5.—Throw the greatest loads at the points of support.

21. ARCHED RIBS.—When under a vertical load distributed on each side of the crown of the arch: To find the total horizontal pressure against the rib below a given point, graphically—

Let c be any point in the rib. In the diagram of strains, draw *oc* parallel to a tangent to rib at c. Draw vertical line *oh* as scale of loads, and take *oh* to represent vertical load on arc AC. Draw *hc* horizontal, cutting *oc* in *c*, then *oc* will be thrust along rib at c, and *hc* will be the horizontal pressure exerted against c B, the part of rib below c.

ARCHED RIBS.

Or algebraically, let H = horizontal pressure or thrust; P = vertical load supported by rib AC; T =

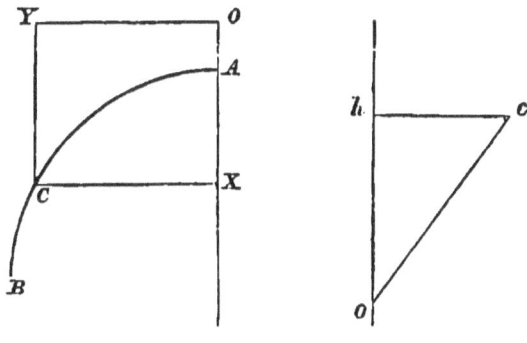

Fig. 39.

thrust along rib at c; also let $ox = x$, and $oy = xc = y$ be the co-ordinates of point c; and let a be inclination of arch at c or oc to the horizon, then—

$$H = P\frac{y}{x} = P \cotan a; \text{ and}$$

$$T = \sqrt{P^2 + H^2} = \sqrt{P^2 + P^2 \cot^2 a} = P\sqrt{1 + \cot^2 a}$$
$$= P \cosec a.$$

At the crown A, the following rule is to be used. To find the thrust at crown of rib: Multiply the radius of curvature at the crown by the vertical load per lineal unit of span there.

RIBS UNDER NORMAL PRESSURE.—In a circular or other rib under a normal pressure, *i.e.* pressure equal in all directions, as in the case of the hydrostatic arch, or an arch sustaining liquid pressure of a given depth, we have the following condition :—

The thrust at any normally pressed point of a rib is the product of the radius of curvature by the intensity of the pressure, the thrust being constant at every point of the rib.

The figure of a hydrostatic arch is identical with an

elastic curve, as that of an uniform spring when bent, and its application to the intrados of an arch-ring is useful.*

Strains found graphically.—As an example of the graphic method of finding the strains on trusses we give the following (fig. 40), as illustrating Professor Maxwell's method of reciprocal figures. Let the load be reduced to vertical loads F_1, F_2, ... on the joints of rafters; G_1 G_2 ... on joints of tie. Let P_1 P_2 be the

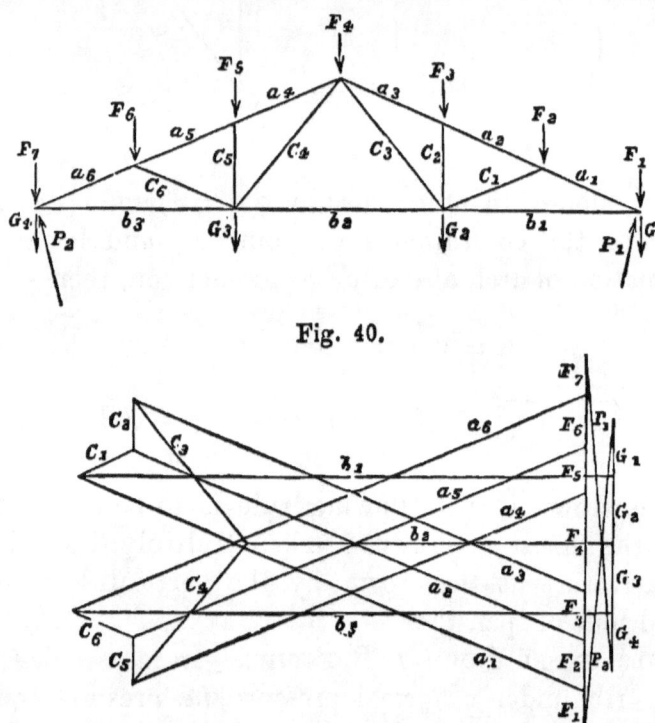

Fig. 40.

Fig. 41.

reactions of walls (shown out of the perpendicular, to avoid overlapping of the lines). To draw the diagram of stresses (fig. 41), the polygon of external forces is first drawn by making F_1, F_2, ... P_2, P_1, G_1, G_2 ... parallel to the corresponding lines in the

* For a full investigation of this subject, see Rankine's "Applied Mechanics," pp. 182, &c.

diagram of truss, and equal to the forces taken in order on any scale. At the right hand joint the forces acting F_1 P_1 G_1, and the stresses on a_1, b_1 at a point are represented in the stress diagram by a polygon. Complete this polygon by drawing lines parallel to a_1, b_1, at the unclosed extremities of lines representing the other forces. At the next joint the forces acting are F_2; and the stresses a_1, a_2, c_1; of these, F_2 and the stress on a_1 are already drawn. Complete polygon by drawing lines parallel to a_2, c_1. This process will give the stress diagram, the lengths of the lines will then give the strains on the corresponding bars.*

22. STRENGTH OF TIMBER.—In fixing the scantlings of timbers of good pine, fir, or oak, the proper sectional area for the compressive and tensile strains should be calculated.

The greatest strain (compressive and tensile) is generally taken at 1,000 lbs. *per square inch*, and the factor of safety averages from 8 to 10. Timber and iron are exposed to these several strains:—

Longitudinal { Extension.
              Compression
Transverse   { Distortion.
               Twisting.
               Bending.

23. JOINTS AND FASTENINGS IN CARPENTRY.—These are of several kinds, and the reader is referred to the works of Tredgold and Nicholson for information of the various joints used.

In designing joints, &c., the following principles should be attended to:—

(1.) To avoid weakening the pieces of timber by cutting or notching too deeply.

(2.) Each abutting joint or surface should be perpendicular to the pressure it receives.

* The letters of reference in the diagrams explain the corresponding strains.

(3.) The area of surface abutted on should be proportioned to the pressure, so that it may not indent.

(4.) The surfaces of contact should be fitted accurately, in order that the pressure should be evenly borne.

(5.) All fastenings, as nails, spikes, bolts, straps, &c., should be of equal strength with the pieces connected.

In lengthening beams, the scarfing should be placed where the bending moment is small. At the compressed side of a beam the two pieces should have a square shoulder; oblique surfaces are bad, and apt to splinter.

The surfaces of a scarf should be parallel to the direction of load, or "up and down," not flatwise.

Notching beams, as a joist or purlin upon a girder or principal, should be effected without cutting deeply into the piece; it is best to cut a square notch in the angles of both, so that the full depth of both pieces remains unimpaired.

Mortises and Tenons.—In mortising a beam the mortise should be cut at the *middle* of the depth of beam, so that the fibres subject to compression and tension at the upper and under side should be left.

A *shouldered* tenon should be used, so that the shoulder should bear upon the edge of the beam, instead of a long and deep tenon, which unduly weakens the beam; the tenon proper should be only 1-6th of the depth of cross beam.

Post and Beam-joints.—When a beam is to rest on the top of a post or posts, a "notch and bridle" joint is to be used, the bridle being an uncut portion in the middle of the bearing beam into which the post is fitted by a groove. The same joint is applicable when a post rests on a beam.

Instead of the ordinary method of framing a king post between ends of rafters, it is best to let rafters abut against each other, and to notch a piece on each side, bolted together to form the suspending pieces (see Tredgold's "Carpentry").

Abutment-joints.—When an oblique strut, as a principal rafter, abuts on the end of a tie-beam, the shoulder and end of strut should be cut at right angles to the pressure or direction of strut, and this abutment should be of sufficient area to transmit the thrust without splintering the timber. Sometimes this is effected by a simple notch on tie-beam, or by a tenon on the strut fitting into a mortise on the tie; or a bridle on the tie fitting into a groove made in the shoulder of strut. Iron straps or bolts are often used to keep the pieces in their places.

Tredgold gives the following rule for finding the length between shoulder and end of beam to prevent the shearing of fibres in the direction of beam. Divide four times the horizontal thrust of rafter in lbs. by breadth in inches multiplied into shearing force of a square inch in lbs. in direction of the fibres, and quotient will give length in inches. The resistance to shearing may be taken at 510 lbs. per square inch; in practice, *four* is the factor of safety.

Rule to find horizontal thrust of inclined beams is given by the formula—

$$\frac{w \times \text{half span}}{2 \text{ Rise}} = \text{thrust.}$$

The above rule for shearing is applicable in ascertaining lengths of tenons from pin-holes, scarfs, and other joints subject to tensile strains.

24. WEIGHT OF FRAMING.—Tredgold gives the following approximate weights:—

	lbs.	lbs.
Weight of a square of partitioning from	1,480 to	2,000
Weight of a square of single joisted flooring without counterflooring.	1,260 ,,	2,000
Weight of a square of framed flooring with counterflooring.	2,500 ,,	4,000

The highest numbers give weight in large buildings and long bearings.

One cwt. per super foot is ample allowance for the probable load on floor of an ordinary dwelling-house, exclusive of floor itself.

Two cwt. per foot is sufficient in ordinary warehouses and factory floors.

25. LOADS ON ROOFS—DATA FOR DESIGN.

	Per square.	Minimum Slope.
Load covering	7 cwt.	4°
Zinc ,,	1½ ,,	4°
Corrugated iron	3 ,,	4°
Slates	7½ to 9 ,,	25½ to 30°
Load for pressure of wind	36 ,,	
Timber framing	5 to 6 ,,	
Boarding ¾ inch	2½ ,,	

The weight of slates per square foot = 8 lbs., tiles 13 lbs., pressure of wind about 8 lbs.

It is usual to take 40 lbs. per square foot as the basis of calculation in designing roofs. In calculating for the weight supported on each truss, the load must be taken from centre to centre of trusses.

Half load rests on each principal rafter, and may be considered as collected at apex and bearings; under this supposition $\frac{1}{4}$ of whole load will be the weight to provide for.

STRENGTH OF MATERIALS.

26. BEAMS OF TIMBER AND IRON.

For a full investigation of the theory of the resistance of beams, the reader is referred to the works of Tredgold, Barlow,

Rankine, and other writers. We shall here content ourselves with a few of the most useful principles and rules.

DEFINITIONS : *Factor of Safety.*—Is the ratio which the breaking load bears to the working load, and varies for different materials.

Dead Load.—Is a steady load, as the weight of a wall.

A Live or Moving Load.—Is a load applied suddenly, and accompanied by change and vibration, as the load which floors and bridges are subject to.

The factors of safety followed in practice, and deduced from experiments made by Mr. Fairbairn and others, are for—

	Dead load.	Live load.
Metals	3	6
Timber	4 to 5	8 to 10
Masonry	4	8

Neutral Axis.—When a beam is subjected to a cross strain, its *upper* fibres are compressed, and its *lower* extended. It follows there is a layer of the material between these portions which is not exposed to either of these strains; this is called the "neutral surface" of the beam. Where this surface or layer is intersected by a vertical plane at section of rupture, the line of intersection is called the "neutral axis" of the beam. The position of the neutral surface varies for different degrees of resistance to compression and tension.

Modulus of Elasticity.—Is the value of stiffness a material possesses within the limits of its elasticity, or it is the strain it can bear without sensible alteration. If l be the original length of a prism of any material, x the change in length produced by a longitudinal force F; E a constant depending on the material, then,

$x : l :: F : E$, or $E = F \dfrac{l}{x}$; E is called the "modulus of elasticity."

Moment of Inertia.—If we conceive a body to consist of a number of heavy particles, and we multiply the mass of each by the square of its perpendicular distance from a given line or axis, the sum of all these products is the "moment of inertia" of the body with respect to that axis. In a rectangle of depth d and breadth b the moment of inertia $= \frac{1}{12} b d^3$.

TENSION: *To find the intensity of direct.*—Divide the load by the sectional area of the bar.

To find the sectional area of a bar to bear a given load.

Divide the load by the proper modulus.

27. RESISTANCE TO COMPRESSION.—This is sensibly equal to the resistance to tension. Direct crushing takes place only when the material compressed offers to the force a length not exceeding a certain ratio to its diameter; when this limit is exceeded, there is a tendency to give way by bulging and bending laterally. The cases included in compression are—

Stone and brick columns of ordinary proportions; pillars and struts of cast iron, when length does not exceed about 5 times diameter; pillars and struts of wrought iron, when length does not exceed 10 times diameter; pillars and struts of dry timber, in which length does not exceed 20 times diameter.

Above rules for tension are applicable to these cases, using the proper modulus of resistance to crushing.

Resistance to Direct Crushing.—Crushing load of material $=$ F $\times$ s, F being resistance of material in lbs. on square inch, and s area of section in square inches.

When the load is not uniform, or its resultant does not coincide with the axis of pillar, the strength of the pillar is diminished in the same ratio as the mean intensity of strain is less than the maximum.

Stone and brick columns should never exceed in height 12 times their least thickness at base.

Long Pillars and Struts.—When the length greatly exceeds the diameter, failure or fracture takes place by cross-breaking.

Mr. Hodgkinson's formulæ for cast-iron cylindrical pillars which yield in this manner are as under:—

Solid Pillars whose length is not less than 30 diameters,

Let d = diameter in inches; L = length in feet; A = constant multiplier, then,

$$\text{Breaking load in tons} = A \frac{d^{3\cdot 6}}{L^{1\cdot 7}}*$$

Hollow Columns.—Let D = external diameter, and d = internal diameter in inches, then breaking load

$$= A \frac{D^{3\cdot 6} - d^{3\cdot 6}}{L^{1\cdot 7}} \text{ when } L \text{ exceeds } 25 D.$$

Values of constant A,

	tons.
For solid pillars, fixed ends	44·16
Hollow pillars, fixed ends	44·3

For short pillars the formula is $w = \dfrac{b \times c}{b + \frac{3}{4} c}$, b = breaking load for long pillars (see above); and c = crushing load for section = 49 tons per square inch.

For Solid Square Red Deal (dry).

$$\text{Breaking weight} = 7\cdot 81 = A \frac{D^4}{L^2}$$

Safe load should not exceed one-tenth breaking load.

Resistance of timber to crushing when dry is from $\frac{1}{2}$ to $\frac{2}{3}$ of its tenacity. Moisture weakens the adhesion of the fibres to about *one-half* of that in dry state.

* $d^{3\cdot 6}$ and $L^{1\cdot 7}$ means that the diameter in inches has to be raised to 3·6th power and the length in feet to the 1·7th power. These powers are readily found by logarithms. (See Table.)

For long solid columns the cylindrical section is found to be rather stronger than the square, and the triangular section stronger than the cylindrical, the areas of section being equal.

28. RESISTANCE TO CROSS STRAIN.—*Moments of Rupture.*—(1.) The moment or strain at any point of a beam fixed at *one* end and loaded at the other, is proportional to the distance from the weight. The resistance of a loaded beam to fracture depends upon the equality of two moments—one the leverage, or Bending Moment, the other the area of section at point of rupture, called Moment of Resistance due to the molecular reaction of the fibres, or when rupture takes place—

$$\text{Breadth} \times \text{depth}^2 \times \text{constant} = \text{Weight} \times \text{leverage}.$$

(2.) When the weight is distributed uniformly, the resultant of weight acts at the middle of the beam, so that the leverage in this case is only half the former, and therefore such a beam uniformly loaded will bear twice as much as when loaded at one end.

(3.) When a beam is supported at *both* ends, and loaded at some *intermediate point*, s, putting P and Q to represent the supporting reactions at ends A and B, the strain at any point s will be same as in first case, or Q × $\overline{\text{B s}}$. If s be the middle of beam the strain will be $\frac{1}{4}$ weight × $\overline{\text{A B}}$.

(4.) When the weight is *distributed*, the strain at middle of beam is only *half* as much as when weight is concentrated there.

The strains at any points may be found graphically by drawing a vertical at the point s, and making it equal strain at that point, and drawing straight lines or curves to ends of beam; these will determine the ordinates at any points. In cases (2) and (4) parabolas determine the ordinates.

29. STIFFNESS.*—Beams supported at ends and loaded in middle. The effect of a load to produce deflection in a beam is, as we have seen (28), as the leverage and load, or the "bending moment." The force producing extension is also as the *length and weight directly*, and the resistance to this force is *inversely as the breadth and square of the depth*. But, combining the equivalents of the extending and resisting forces, deflection is as the *weight and cube of length directly*, and as the *breadth and cube of the depth inversely*, that is, the *resistance* to deflection is directly as the breadth into the cube of depth, and inversely as the weight into the cube of length.

Let L = length of bearing in feet; W = weight in lbs.; B = breadth in inches; D = depth in inches; a = a constant determined by experiment for the material; s = deflection in inches then—

$$s = \frac{L^3 \times W \times a}{B \times D^3} \text{ and } a = \frac{40 \times B \times D^3 \times s}{L^3 \times W} \quad . \quad (1)$$

Deflection should not exceed 1-40th of an inch for every foot in length, therefore dividing above formula by 40

$$s = \frac{L^3 \times W \times a}{40 \times B \times D^3} \quad . \quad . \quad (2)$$

Tredgold gives the following rules for the stiffness of beams:—

To find the scantling of a piece of timber to sustain a given weight when supported at ends.

CASE 1. When *breadth* is given.

Rule.—Multiply the square of length in feet by the weight in lbs., and this product by value of a. Divide product by breadth in inches, and cube root of quotient will give depth in inches.

* The following are the most useful rules for the architect, as the undue deflection of beams would seriously affect a structure.

Case 2. When *depth* is given.

Rule.—Multiply the square of length in feet by weight in lbs., and multiply this product by value of a. Divide last product by cube of depth in inches, and quotient will be required breadth.

It will be obvious a certain proportion between depth and breadth is desirable to give a maximum rigidity. Tredgold gives this when the breadth is to depth as 0·6 is to 1; thus $\frac{L}{\sqrt{D}} \times 0\cdot 6 =$ least breadth required to prevent beam yielding sideways. The stiffest beam that can be cut out of a round tree gives breadth to depth as $1 : \sqrt{3}$ or $\cdot 58 : 1$.

STRENGTH OF BEAMS.—The weight a beam will carry without fracture depends upon the relation which exists between the moments of "*rupture*" (28) and "*resistance*," or the equality of the moments. The moment of resistance is evidently due to the sum of forces of the fibres of the beam at section of rupture, or a compound action of resistance to tearing the *lower* fibres asunder and crushing the upper fibres, the "neutral axis" being the limiting extent of each action. Hence the resistance of a beam at any section is directly as the moment of inertia (26) about the neutral axis of the section, and inversely as the distance of that axis from the farthest edge of the section, or $\frac{c\,b\,d^2}{6}$. The formula is reduced to a simple one where the length of beam is given in feet, the other dimensions being in inches, or

$$\frac{B \times D^2}{L} \times c = \text{breaking weight; and}$$

$$\frac{B \times D^2}{L \cos \phi} \times c \text{ when beam is inclined.}$$

Values of constants a and c in above formulæ are for the timber ordinarily used—

	Cohes. force Per sq. inch. lbs.	Stiffness (a)	Strength (c)
Fir, Riga . . .	12,600	·0114	359
„ Memel . . .	„	·0089	545
Oak, English . .	12,000	·0119	557
Pine, Red . . .	10,000	·0148	447
„ Yellow . . .	„	·019	383

(The above constants are based on experiments made by most eminent authorities.)

When a beam is fixed at one end and loaded at the other the breaking weight as found above $= \dfrac{w}{4}$; when weight is uniformly distributed it $= 2$ w; when beam is fixed at both ends and loaded in middle, breaking weight $= 1\tfrac{1}{2}$ w.

30. FLOORS AND ROOFS.—To find the depth of a joist when length of bearing and breadth are given, the distance apart being 12 inches from centre to centre.

Rule.—Divide the square of the length in feet by the breadth in inches, and cube root of quotient multiplied by 2·2 for fir will give depth in inches.

GIRDERS FOR HOUSE FLOORS *ten feet apart*.

Rule.—Divide square of length in feet by breadth in inches, and cube root of quotient × by 4·2 for fir will give depth in inches.

FLITCHED BEAMS.—Wooden beams with flitches of wrought-iron plate slightly increase the strength.

Hurst gives the following formula:—

B = breadth in inches.
D = depth in inches.
t = thickness of iron flitch in inches.
L = bearing in feet.
W = breaking weight in centre, in cwts.
C = constant = 3·662 for oak; 3·024 for Baltic fir.

$$W = \frac{D^3}{L}(c_B + 30\,t).$$

One-twelfth is assumed as thickness of flitch.

31. PRINCIPAL RAFTERS.—To find scantling when there is a king post, length, breadth, and span in feet being given

$$\cdot 096 \times \frac{L^2 \times \text{span}}{B^3} = \text{depth in inches.}$$

When there are two queen posts multiply by decimal 0·155 instead.

COMMON RAFTERS.—Depth is found by following rule, breadth being 2 to $2\frac{1}{2}$ inches.

$$0\cdot 72 \times \frac{L}{\sqrt[3]{B}} = \text{depth in inches.}$$

PURLINS with long bearings may be trussed with wrought-iron rods. Depth may be 1-12th the span, and distance of stays apart 1-3rd span.

SECTION III.

32. IRON CONSTRUCTION.—The same principles and formulæ which determine the stability of timber frames are applicable to iron structures. The young architect should make himself thoroughly conversant with the experimental researches of Tredgold, Barlow, Hodgkinson, Fairbairn, Kirkaldy, and others who have investigated the principles of iron structures, as this material must ultimately take the place of timber in all important structures. The neglect of iron among architects has given engineers a great start in the art of construction; and the architect who wishes to keep pace with modern invention and science must jealously guard all encroachments upon those materials which are reckoned among the great resources of the present age.

Among the many uses for which iron may be employed we may enumerate flooring, roofing, either partially or wholly, supports as columns, bracketed structures, stairs, window-casements, besides the multifarious purposes where rigidity and strength are required in the least space.

Cast iron affords the architect a material which he can readily adapt in innumerable cases in which timber or stone is inadmissible, as in balustrades, crestings, spandrels, brackets, verandahs, balconies, &c., &c. It can be cast or moulded into every conceivable form or pattern, as in moulded gutterings, pillars, &c.

Wrought iron, though less plastic, is even more valuable as a constructive material, and may be employed in all cases where heavy weights are to be carried, or light structures, as roofs, erected. For girders, simple or trussed, domical construction, and where extreme lightness and rigidity are to be combined, no material, or mode of construction, can supersede it. As a fireproof material it is preferable to cast iron; and when cased in concrete, or plaster, or protected from the direct action of fire, it may ultimately be made the most resistant of all materials.

CAST IRON should be limited to the action of stationary loads. Its brittleness, from the carbon it possesses, renders it necessary to confine its use to purposes where great changes of temperature are not felt. Its crushing resistance is about *six times* as great as its resistance to cross strain; hence it should be used only for beams, columns, struts of short length, sockets, and castings which are not exposed to sudden cross strains.

The best kinds of cast iron for large structures are Nos. 2 and 3 of grey cast iron; granular white cast iron is the hardest, but more brittle.

Iron is sounder by being cast under pressure and in a vertical position; the air bubbles ascend to the head. Allowance should be made for expansion and contraction of the castings. In designing patterns for castings all abrupt variations in the thickness of metal should be avoided, so as to prevent unequal contraction in cooling, and thereby injury or fracture of the iron.

STRENGTH OF CAST-IRON BEAMS.—From experiments made by Prof. Hodgkinson it has been found that the strongest form of section for a beam of given depth is that in which the top and bottom flanges are to one another in the same proportion as the ratio of resistance of extension to crushing, or as 1 to 6.

For the strength of an iron beam we have the following approximate formula; putting a for sectional area, in inches, of bottom flange, d for depth in inches, l for bearing in inches, w breaking weight in centre, in tons, and c, a constant, we have—

$$w = c\frac{a\,d}{l}$$

The constant generally taken in practice is 25·0 tons. The central web between flanges should be made to taper upwards, and the lower flange be made to contain the most metal, its width being three times, or twice, as great as the upper flange, since the latter is subject to compression and the lower to tension. A strain of more than $1\frac{3}{4}$ ton per inch is injurious on lower flange. The depth of cast-iron beams should be from 1-12th to 1-16th of the span.

For a girder of *uniform strength* in every part of its length, it may be diminished towards the ends in the proportion of the rectangles of the segments at every point. Thus if depth at ends equals $\frac{2}{3}$ of centre depth, an elliptical curve may be drawn for the web and upper flange. Or a similar diminution may be made in the *plan* of lower flange by making it a double parabola, or of segmental curves. It is evident a uniform section throughout entails a waste of metal at the extreme ends.*

WROUGHT-IRON BEAMS.—Wrought, or malleable, iron is made by the abstraction of carbon and other impurities from pig iron. It is more fibrous, tougher in quality, and more compact than cast iron. Its resistance to compression is about *two-thirds* of its resistance to extension, and its tenacity recommends it for resisting cross and tensile strains. Kirkaldy gives 25 tons per square inch for the tensile strength of wrought iron in bars, and 22 tons for plate iron. In designing pieces of forged iron subject to sudden shocks, angles and irregularities should be avoided, as they induce fracture.

For *Wrought-iron Plate Girders* the following formula is approximately correct.

Let l = length of girder in feet, w = distributed load in tons; d = depth in feet; s = strain on top and bottom flange, at centre, in tons. Then

$$s = \frac{w\,l}{8\,d}; \text{ and when total load is in centre } \frac{w\,l}{4\,d}.$$

Wrought iron may be strained in compression to four tons per square inch, and in tension to five tons; so the effective sectional areas of top and bottom flanges should be determined accordingly, the rivet holes being taken into account.

Depth taken at 1-12th span is regarded as economical for a straight girder.

---

* Cast-iron beams should not be loaded with a permanent load of more than one-sixth of breaking weight.

The "box" girder is found to be stronger than the I-shaped of the same weight, in the proportion of 100 to 93. For large girders the "lattice" principle is most economical; it exposes less surface to wind than the solid web.

In large plate girders the web must be stiffened at intervals by angle irons and stay ribs.* Width of bottom flange may be taken about $\frac{2}{3}$ depth. The top flange is generally made of larger sectional area than the lower, to provide for the comparative weakness of wrought iron under compression. Brunel ingeniously effected this by giving the top flange a curved shape without increasing area of metal.

(For detailed particulars the reader is referred to Hodgkinson's work, and other treatises on wrought-iron plate girders.)

ROLLED IRON BEAMS, or joists, are now generally used for ordinary floors and short bearings. The metal is rolled in one piece. Sometimes, for additional strength, two rolled beams are bolted *one on the other*, or horizontally combined by additional flange-plates, which constitutes a very effective beam. Steel is now often used where great tenacity is required, or the action of fire is to be withstood.†

Dr. Fairbairn has, in a series of experiments, found that the effects of long-continued vibratory action, and changes of load upon wrought-iron bridges and girders seriously affect the material, and that five tons per square inch should be the *maximum* strain produced by both permanent and moving load.

---

* The web or part between flanges has to resist the shearing force. When a beam supported at ends is loaded in centre, the shearing force at every point $= \frac{1}{2} w$; when load is distributed it is greatest at supports, and nothing in centre.

† Iron beams and columns should be cased in earthenware or plaster as a non-conductor. These casings could be moulded to any design. Plaster hollow pots between joists ceiled underneath are equally fire-resisting.

The chemical action induced by moisture and smoke on iron renders it necessary to provide a due thickness of metal in all permanent structures, as well as the necessity of a coating of paint, which shall arrest corrosion.*

STEEL, as a building material, is now, for special purposes of construction, superseding malleable iron. It is compounded of pure iron and carbon; the proportion of the latter varying according to the requirements for ductility or hardness. Bessemer's process has greatly facilitated the manufacture, and it can be rolled or hammered into joists, railway bars, &c.

Girders, and joists, and pillars made of steel are less liable to fracture or twist than iron under the action of fire. For roof framing, too, it offers extreme lightness and durability. Its resistance to tension is only half its resistance to compression; so the lower flanges of girders should be twice the area of top. The strength of steel columns is said to be double that of solid cast-iron ones of equal size, and about 30 diameters. For steam boilers steel plates of $\frac{5}{16}$ inch will bear a pressure of 100 lbs. on the square inch. Mr. Kirkaldy's experiments have shown that a high breaking strain *may* be due to hard unyielding quality, and a low one to softness. The contraction of area at fracture is an essential element in judging of quality. It appears also that the breaking strain of puddled steel plates and iron is greatest in the direction in which they have been rolled; in cast steel, the reverse.

---

* See Mr. Matheson's "Works in Iron." The author recommends oxide of iron paints for all iron work, and bituminous paint for the inside of pipes. Barff's process of exposing iron to superheated steam has been well spoken of.

PART V.

SANITARY CONSTRUCTION.

THE questions involved in the Drainage, Ventilation, and Warming of buildings demand the serious attention of the young architect. Next to the stability and arrangement of his structures, their sanitary condition is the most important of the various functions he is called upon to provide for. Here an extended knowledge of physical science, as regards the phenomena of heat and atmospheric changes, the origin, distribution, weight, and elasticity of gases, the chemical properties of air and water, the principles of hydrostatics, pneumatics, and thermodynamics will be of infinite service to him. He should remember that his profession entails upon him more direct and responsible duties than those of design in the abstract. He really becomes the administrator of public health or disease. Whole families depend on his careful attention to matters of drainage and ventilation, and public health is in a large degree entrusted to his control in the design and construction of large public buildings, as workhouses, schools, factories, churches, and the like, where human beings congregate in large numbers, and disease and death itself are frequently propagated by infection or otherwise.

Section I.—Warming.

To regulate temperature is one of the greatest secrets of the art of warming. The laws of heat belong to a special branch of natural philosophy, to which the student is referred. Combustion is the chief source of artificial heat, and consists of the rapid union of the oxygen of the air with various substances for which it has a strong chemical attraction. Our ordinary combustibles or fuels are composed chiefly of two simple elements, carbon and hydrogen. The carbon of the fuel unites with a certain proportion ($2\frac{2}{3}$ times) of its weight of oxygen, and forms carbonic acid gas, and the hydrogen unites with about 8 times its weight, forming water or vapour. The complete combustion of a pound of coals requires about 230 cubic feet of air. The nitrogen of this air, which forms four-fifths of its bulk, mingling with the carbonic acid and vapour, ascends in a gaseous form from the fire. Smoke from coal is the vapour which rises from the carbon and hydrogen or bitumen in it when the heat is only about $600°$, a greater heat constituting carburetted hydrogen, or coal gas.

Open Fireplaces.—The open English fireplace, with its air of cheerfulness and comfort, has a claim upon our social life which it would be difficult to displace. It is, however, with all its advantages, very wasteful of fuel. It is estimated that about *one-half* of the heat produced ascends with the smoke and is wasted. About a fourth part of the heat which is radiated is lost also in the space between the fire and mantel; indeed, Dr. Arnott calculated that only about *one-eighth* part of the power of the fuel is realised, the rest being dissipated. Another great objection to it is the creation of draughts. Various improvements have been made to obviate these evils, first, by diminishing the conductive power of the metal. Fire-bricks are now generally used in the backs and sides of grates to radiate the heat into the room, instead of metal which conducts the heat to the walls; secondly, by forming the sides or covings of the chimney-mouth in such a manner as to throw the heat into the room, and to impede its entrance into the chimney, a still greater advantage is attained. Thus, instead of the fire recess being square, the sides should be inclined to the back at an angle of about 130°. Fire-brick slabs moulded in ribs are often placed in these positions in lieu of metal plates polished, and are preferable in retaining and throwing *out* the heat. Plates of rough metal being good conductors of heat, absorb it by passing through to other materials the heat they receive, and hence they lose heat instead of retaining it. Radiant heat does not affect the surrounding air; the rays of heat pass through without directly warming the room; the warmth afterwards felt is due to the reflection and absorption of the rays of heat by the surrounding objects—the walls, floor, furniture, &c., which being in contact with the air give out their heat. It is thus explained how a large room or massive stone building requires a considerable time before its chilling effect is subdued by heat, a great quantity of such heat being lost in the transmission of it from one body to the other. Hence cold glass and stone surfaces take a considerable time before they are sufficiently warmed to throw off the required amount of heat, and during this time strike chills to persons near them.

Grates should on this account present a large surface of heat in front, surface rather than depth of fire is desirable. The shape of fire-box or grate should be designed in reference to these principles, and many modern grates attain the desired effect in a certain measure.

Again, the chimney-throat or mouth for the exit of the products of combustion should be only of sufficient capacity to allow these to pass, instead of a large space which is continually drawing in the warm air of the room, and creating draughts from doors and windows towards the fireplace. " Register" stoves answer this end to a certain degree, and Dr. Arnott's smokeless grate was constructed to meet it. Many of the recent attempts to improve fire-grates are really based on the principle he adopted. Dr. Arnott's grate has no bottom, but is fed by a box open at top, into which the coal or fuel is placed (about 20 or 30 lbs. per day). The wood is ignited at the top of the fuel, a layer of cinders being above this. The ignition of the wood ignites both the coal below and the cinders above, the pitchy vapour from the fresh coal rising through the wood, flame, and cinders. When the cinder is once ignited, the rising bitumen is burnt. The fuel as it burns is raised by a lever acting on a false bottom to the box. The valuable quality of this fire is its tenacity to burn without attention. The draught can also be regulated by a small slide at the bottom of coal-box, which also forms the outlet for the ashes.

The Arnott grate is also provided with a cover or hood of metal over the fire-grate to prevent the egress of the warm air of the room up the flue, the only air being admitted through the fire in front. The hood is furnished with a throttle-valve or damper, having an external index, so as to further regulate the current. A saving of from one-third to one-half of the fuel is claimed for this grate by the inventor.

Instead of the metal hood, however, a simpler means of reducing the chimney-mouth is by contracting the brick throat of the chimney. It is evident that this grate combines the requirements for regulating the temperature of a room as far as the fuel is concerned. Various means have more recently been devised for economizing fuel and utilising the heat of open fire-grates. The principles upon which nearly the whole of these inventions are based may be summed up as follows :—

1. An open projecting fire and large heating surface.
2. Fire-brick sides and back.
3. Contracted chimney opening to avoid the rapid escape of the warm air of room into flue.
4. A separate and distinct ingress for the cold air to feed the combustion and to prevent draughts from other sources.
5. Hot-air chambers behind and at the sides of the fire, communicating by open gratings or valves in front, or over grate.
6. Consumption of smoke.

The "Wharncliffe" grate, Captain Galton's stove, the stoves patented by Messrs. Shillito and Shorland, and other patent fire economizers, more or less meet the requirements we have alluded to, though they cannot claim the originality asserted, but are all modifications of the inventions introduced by Cardinal Polignac, Count Rumford, and others in the beginning of the last century, and the later stoves of Sylvester, &c.*

It will be seen, then, that upon the construction no less than the materials used to radiate the heat of the fire depends the economical distribution of heat; that direct radiation is wasteful of fuel, and that the desired object is to be attained by increasing the surface of

---

* The openings or inlets in all these grates should be placed high, or the warmed air would find its way into the chimney; especially is this the case where no proper outlets are provided.

metal or fire-brick so as to diffuse the heat of the fire both by radiation and conduction; to prevent conduction of heat to walls or its current up the smoke flue, which can best be done by hot-air chambers, with or without gills; or by a combination of the two principles, a warm-air stove with an open fire. The patents recently brought out, especially those noticed above, seem to fulfil these desiderata, and are worthy the attentive study of the architect.

Warming by close stoves, though a preferable mode to the simple open fire, is open to objections; the metal becomes so heated as to scorch the air and dust in contact, and render it unwholesome.* It may be stated that air to be wholesome requires a certain proportion of vapour; a cubic foot of air heated to 80° is capable of absorbing five times as much moisture as the same quantity at 32°. Thus it becomes necessary to supply to such stoves a water-pan, so that the heated air should be moistened by the evaporation. A stove in which the fire-box is surrounded by brick or cased so as to allow an intervening chamber of hot air is preferable, and the heating surface is thus increased in proportion.

Gills placed within the two metal cases, or radiating plates, as in the Gurney stove, may further be employed in increasing the heating surface. The "Cockle" stove, Dr. Arnott's, and various modifications of the same principle, too numerous to mention here, placed in basement, may be used in the heating of halls, public buildings, &c., and the warm air be conveyed by flues to different parts.

Whatever system of warming be employed, it is

---

* Objections to water-pipes heated on the high-pressure system have also been raised for similar reasons, though we think for large systems of heating it answers.

necessary that the fresh cold air should be admitted in a gentle and diffused manner, and not in sudden currents from chinks and crevices of doors, &c., as is generally the case, giving rise to draughts and the concomitant results, colds, influenzas, and the sharper pangs of nerve pain. This can be effected by allowing the fresh air to enter through some channel or flue into the hot-air chamber heated by fire-gills, or coils of pipe, as the case may be, and, having become heated in its passage, allowed to enter the apartment through perforations in the stove or around it in an equable flow of fresh warm air; or otherwise be conducted through flues in the skirtings, &c., to various parts of the room.* The admission of external air may easily be regulated by valves or "hit and miss" slides.

Conservation of Heat.—Warming to be effectually applied to buildings requires not only the distribution of heat by fire-places, stoves, hot water, or other apparatus, but such a construction of the building as will prevent the undue waste or escape of the heat generated. Indeed, we consider the conservation of heat or warmth the secret of the art.

Hollow Walls.—To this end all walls should be as *retentive* as possible, that is, should be built of non-conducting materials. As brick or stone is the ordinary material employed, it is, we think, desirable to construct walls, especially in cold or exposed situations, with a hollow, or in two thicknesses, a plan frequently adopted in southern parts of England. The layer or stratum of air between the outer and inner parts becomes an effectual barrier to the passage of the heat, and forms a non-conducting medium of great value in equalising the interior temperature under extreme changes. Damp walls absorb heat to a great extent

* This last method the writer suggests as preferable.

by the evaporation of moisture from their surface, but by having hollow walls perfect dryness is insured. Even iron walls built on the cellular plan with a lining or plaster may be made impervious to heat or cold. Cold walls, the surfaces of windows, floors, &c., create currents of air which rob the heat that should be made available.

Windows.—These become a great source of waste of heat. Besides the currents of air ordinary sash and casement windows allow to pass through crevices and imperfect fitting, their cold glass surfaces absorb the heat, or allow it to pass by conduction. Currents of cold air are continually formed, and descending on the heads of those who sit near windows, are a continual source of discomfort in houses and public rooms. To avoid this waste of heat, double windows may be employed in all cold situations; they also, near streets and roads, shut out the noise and dust. Double sashes may be used, or two sheets of glass in the same sash or casement frame may be adopted with great advantage.

Doors.—For the same object double doors should always be placed at entrances or lobbies, and in every passage subject to cold currents.

Roofs should have felt, plaster pugging, sawdust, or some non-conducting material inserted under the slates, or between the rafters. The advantages of such roofs are great. They retain heat in cold weather, and effectually resist the penetration or conduction of heat in summer-time.

Section II.—Ventilation.

The principles of heating and retaining warmth and a free ventilation are frequently irreconcilable, and the means adopted to ensure the one often counteract those employed for the second object.

Circulation of air.—The *circulation* of pure air at a certain degree of warmth is the desideratum. If an efficient system of heating be adopted without adequate means for the escape of the products of respiration or combustion, a house becomes an oven; or if free inlets are provided through the apertures of doors and windows, intolerable draughts render it uninhabitable. But if the inlets are so arranged as to avoid the inconvenience of direct down or cross draughts; or if the pure air is first warmed by passing through hot-air chambers, hot-water coils, or round stoves before its admission to the apartments, and, further, a sufficient means of escape be provided for it after it has been respired, the object of ventilation is attained. A continual circulation of one part of a house with the other is requisite, and the best method of effecting this desirable object is still a vexed question among architects. Captain Galton's ventilating fire-place, adopted in barracks and hospitals, to some degree effects the object as regards ordinary rooms.* The warm-air currents are admitted high up in the chimney breast. The inventor claims for his method an equable temperature in all parts of the room; that it saves *one-third* the fuel, is free from smoke, and prevents down-draughts if placed with an unlighted fire in same room. With properly regulated inlets for the fresh air behind the grate, this means would afford all that was necessary for warming. The foul air must, however, be drawn off. This may be accomplished by valved outlets, as Dr. Arnott's, entering a fire flue, or, better still, a flue purposely made for this object, either combined with or near the smoke flue. We think the cornice of a room admirably lends

---

* Various other ventilating stoves, as the " Manchester," Boyd's, Pierce's, Steer's, Langdon, have recently been introduced, and are more or less successful approaches.

itself to this purpose if made "hollow" connecting with a flue. Sun-burners are efficient extractors when lighted.

The extraction of the foul air at the floor level, advocated by some theorists, is, we think, objectionable, and unsound in principle; as it is a law of physics that a gas expands and ascends when heated, the colder strata of air taking its place, we therefore think a system of ventilation which follows this natural order the best, and the ceiling level is for this reason the place for the exit of the vitiated air. The carbonic acid gas mixed with the air does not by its greater weight separate and fall to the lower level, as imagined by many writers, but tends by the law of diffusion of gases to diffuse itself throughout the room. To increase the circulation of this upward current in flues without fire, the draught may be maintained by gas-jets or the flues warmed by other means. By the laws of hydrostatics, there is always a column of ascending air in a chimney or warm flue, the heavier column of colder air outside tending to take its place.* Hence, a fire-place offers the best means for perfect ventilation, and should be adopted in preference to openings in windows, &c., through which currents of cold air are continually entering, especially if no special admission is afforded at lower levels; and for the ingress of fresh air we would arrange openings either in the floor or through skirtings which may be warmed by flues or pipes in large buildings; or in the case of small apartments, inlets at the fire-grates, as noticed before. The "Manchester grate" is one of the best means yet devised of warming and ventilating several rooms by the waste heat of a single fire, and this is

* The draught or velocity of air in chimneys is found by formula:— Velocity in feet per second $= \cdot 365\sqrt{H\,(T-t)}$; H being height of shaft n feet; T temperature of room; and t ditto of external atmosphere.

effected by warm-air flues of metal being carried up from the hot-air chambers surrounding the fire-box. By these warm-air flues good ventilation can be secured by valves or gratings in all the rooms so warmed.

But another mode of making the circulation complete in a house, is by converting the inner halls or staircases into warm-air chambers by stoves or hot-water coils, and thus diffusing through the rooms which open from those halls an equable flow of fresh warm air.*

By thus confining the heating apparatus to the halls in the first instance, the necessary supply of cold air is warmed before admission into the apartments, and the evils of draughty rooms averted, as there will not then exist that continual rushing in of air to supply the fire-places under this system. We have adopted this system with success in some cases, and great economy of fuel is the result.

It will thus be seen a perfect system of ventilation is inseparable from warming, and both objects may be effected at the same time. It is the neglect of the mutual relation between the two objects which has been instrumental in impeding the advance of this great hygienic branch of construction.

The great principle may be shortly stated to be to provide means for the entrance of fresh air at a warm temperature and at a low level, and its extraction at a higher. Its removal is in proportion to the warmth of the apartment compared to the external air, if the natural system is adopted; or if heat is applied to accelerate, it will depend on the degree of heat, the law of equilibrium between the cold and warmer

---

* Pipes or channels behind the skirting may be provided for the passage of the warm air. A diffusion of warmed air from a basement chamber through proper flues opening into halls and rooms I believe to be the most economical system of warming large buildings.

atmospheres always tending to create a pressure towards the vacuum. So long as the incoming currents are rendered warm and imperceptible by entering through small and well-distributed apertures, the rapid circulation is harmless and promotive of health, and the evils of a strained, unprovided admission through crevices is perhaps more destructive to health than the want of egress. In summer-time these effects are not so apparent, though the two operations of ventilation—the removal of the foul air and the admission of the fresh—are still required.*

It may be useful to note that the respiration of one individual requires at least 500 cubic feet of air per hour; † and every cubic foot of gas requires 10 cubic feet of air and produces about 1 foot of carbonic acid.

In unions, hospitals, prisons, &c., it is usual to allow for day-rooms about 300 cubic feet of air to each inmate, and 1,000 or more cubic feet in the dormitories.

An excellent method of extracting the vitiated air of hospital wards, public rooms, &c., is worth the attention of the architect. Its action depends on the suction created in a tube if it be blown across at one end. The principle has been applied to ventilating tubes carried up from the upper part of rooms. Deflecting plates are fixed at the upper end at an angle, and the horizontal currents of air produce a partial vacuum at the top, and cause an up draught to take place. Tredgold first introduced the plan, but recent inventors have improved upon it by using vertical corrugated plates or horizontal ones; such are the patents known as Boyle's air-pump ventilator, and Messrs. Banner's cowls. These can be made to suit buildings in any style, and the system has

---

* In warm weather, the current of air is often reversed, as the air is more rarified outside.

† Dr. Arnott thinks 20 cubic feet per minute necessary in soldiers' sleeping-rooms. In houses 600 cubic feet of space per adult should be allowed.

been applied to ridges and dormers of roofs, as well as to the ventilation of soil pipes and drains. Sheringham's ventilators are frequently used with success.

Section III.—House Drainage, &c.

As the object of these remarks is chiefly confined to the consideration of drainage as it affects our houses, it will be unnecessary to enter into any detail of the main question of sewage other than its bearing on this subject.

General Sewerage: Gases.—There is doubtless much room for improvement here; the pressing evils of the present system seem to me easily averted by proper means of provision in connection with our house drains, and I allude chiefly to large town systems of drainage, where it is more than ever necessary to exercise precaution. Especially is this necessary in low-lying districts and seaport towns, where sewers and drains frequently become tide or "water locked" during several hours every day, till the tide allows the escape of the pent-up sewage. Under the diluent system of disposing of our sewage—and this is at present the only practicable system of drainage available in large towns—there must ever be a large though varying volume of gas generated in the sewers, and pent up within them with no means of escape or relief, except through imperfect gullies, man-holes, and house-traps and drains. Through such vents this gas sometimes escapes, often to the injury of the public health, but especially to the occupants of houses. The pressure of this gas is very powerful at certain times, either from the displacement caused by the sewage water, heavy rainfalls, or by the temperature of the sewers. Now it is evident that the more air-tight and perfect the traps and pipes are that prevent the escape of this noxious gas, the more compressed and concentrated is the gas, and the more eagerly it finds a vent through a weakly guarded passage or trap. Again, the modern system of draining our houses invites the admission of these poisonous gases in two ways. By virtue of the greater lightness of these gases (as nitrogen, sulphuretted hydrogen, ammonia, &c.), increased by the ascensional force created by the higher temperature of our houses, the cold drains empty their dangerous product into our water-closets and through our sinks, &c. Secondly, the *arrangement* of the pipes within our houses is another and perhaps greater source of evil. Our soil-pipes, for example, are generally in an upright position, and, stupidly enough, are by the present mode of construction actually stopped under our very noses by a *water* check, or "trap," as it is called, which is placed under every closet-pan, sink, &c. This water check, or seal, which is generally ineffectual, and often inoperative from various causes, becomes useless every time the closet action takes place, the downrush of matter allowing the free escape of foul gas through it. What a pitiable piece of human ingenuity, that permits admission to the most dangerous foe every time our closets are called into use!

REMEDIES.—Now what are the remedies proposed or existing for this crying evil, the most dangerous and deadly of our modern sanitary regulations? They may be summarised into three classes:—1st. Those remedies that *remove* the *cause* of mischief; 2ndly, those which deal with the main system of sewers and their outlets in relieving them of the gas, &c.; 3rdly, those remedies which may be applied to house-drains. They are either chemical or mechanical in operation.

Now one of the most important examples of the first class is that known as " Moule's Patent Earth Closet

System," in which, instead of water used as the flushing and cleansing agent, *dry earth* or ashes are employed to absorb the noxious portion of excrementary matter and deodorise it. In country districts where no water system exists, or where ready means of utilisation of the product is at hand, this disinfecting process is both valuable and economical; but in large towns I do not see, under present circumstances, that it can come into general operation.*

Several plans exist or are proposed under the second class. These embrace various schemes for the disinfection and the deodorisation of sewage matter by chemical agencies. Charcoal, chloride of calcium, lime, and other substances, either to absorb or prevent the generation of ammonia, to destroy the products of decomposition, or to act as precipitants (as the alum, blood, and clay of the A B C process), are used for the deodorisation and utilisation of the sewage matter.

Other plans employ mechanical means of sewer ventilation either by the erection of ventilating shafts, or by using the up-cast shafts of factories as extractors. The latter mode is perhaps the most direct and practical for the disposal of these sewer gases, although attended with some difficulty. There can be no question as to the desirability of ventilating our sewers, along the main arteries or at the highest and lowest points; and such means, in conjunction with a general system of house-drain ventilation, as will be next discussed, cannot fail to have a marked influence on the health of large populations. The noxious effluvia, however, arising from man-holes and entrances to sewers in our public streets, is often detrimental, and proper means of egress or disinfection of the gas should

* Carbonised refuse, or charcoal, is a good absorbent and deodorant, and has been adopted lately with the " pail system."

in all cases be provided: some charcoal filters get saturated and do the latter imperfectly.*

The third class of remedies deals with the evil at our houses or through our house drains. Several plans have been suggested, all more or less depending on suction or on the natural tendency of such gases to rise. Some propose that the flues of houses should be made up-cast shafts; others that rain-water pipes should connect with the drains, and thus carry away the noxious effluvia above our roofs. Again, the most general remedy adopted is that of carrying ventilating pipes from the upper ends of our soil-pipes or closet syphon-traps to the roofs, or connect them with our stacks or fire flues. The last remedy is perhaps the most practicable of the kind, answers the purpose tolerably well, and can be readily applied.† A still more efficient plan, however, remains to be noticed— that of intercepting or giving free escape to the sewer gas before entering the house drains. This may be done by having a kind of cesspool or shaft outside the house open to the air, and communicating with the trap or syphon between sewer and house. Such a plan is proposed by Professor Reynolds, and is one of the best and simplest means yet devised. I think this plan one of the best, inasmuch as it cuts off all gas or effluvia communication with the sewer *outside* our houses instead of dealing with it after finding a channel for its escape inside; for it must be observed, there is little gas, comparatively, generated within our house drains, the products of decomposition not having time to become very noxious. It may be observed here also that this class of house remedy is more thorough than

* Mr. Baldwin Latham's charcoal trays fixed at the summits of manholes or shafts have been used advantageously when a sufficient number is employed.
† Many of these plans have been adopted with more or less success.

any of the sewer remedies, and for this reason,—that in a large system of sewerage, the house drains and outlets become naturally, by virtue of the laws of pneumatics, the only places for egress of the pent-up sewer air.

VENTILATING JUNCTIONS.—A still more efficient and ready means of disengaging the sewer gas as well as any effluvia that is generated within the house drains, I now bring before my readers. It is based on the principle above enunciated—that of intercepting the sewer gas and giving it free escape externally before entering our houses. Now all flue and pipe ventilators, however fixed, have the objection that they are costly and complicated, and often not so effectual as could be wished, by reason of their length. The invention or method which I propose is free from these objections, and the simplicity of it, and the variety of ways in which it can be applied, claim for it attention from all sanitary authorities, boards, urban or rural, medical and scientific men, and especially architects, builders, and house plumbers. I have called it the "Ventilating or open Junction" because this designation seems to convey best its use and application. Its simplest form may be described as a small trough open to the external air and inserted between the soil or waste-pipe and the syphon or other trap of the closet or sink. Its more perfect or developed form combines the syphon-trap and ventilator, and may further be provided with a tray for charcoal or some disinfecting substance through which the deleterious properties of the gas are destroyed. The last alternative, however, is not necessary under ordinary circumstances, for as the escape of gas is *constant* and not intermittent, as in ordinary closed soil-pipes, there is no time for the generation of the noxious products of decomposition.

It is evident any ordinary water-closet or soil-pipe

can be provided with a ventilator by simply cutting out a portion of the pipe under the trap, and introducing the open trough or gas outlet, which is simply a trap open to the air.

It is a well-known fact, and recently brought before the notice of the British Medical Association by Dr. Andrew Fergus, that lead soil-pipes, where they are nearly horizontally placed, and the upper bends of syphon-traps of the same metal, become under the corrosive action of the gas evolved from decomposing excretal matter, literally eaten through. Several specimens of decayed leaden pipes were exhibited, showing that the lower half of the pipes, being in contact with the soil and liquid, were not affected. In all cases which Dr. Fergus attended for diphtheria or typhoid, he had found escape of sewer gas from soil-pipes. Such evidence shows that the egress for gas should be ample, and be at the uppermost points, and be provided as often as possible in the length of the pipes. All bends and unnecessary elongations should be avoided, hence small ventilating-pipes, by their friction, retarding the gas, are ineffectual. A ventilating-pipe should never be less than an inch and half in bore.

It may possibly be objected against the "ventilating trough outlet" that it allows escape of sewer gas to take place at inconvenient points, which may be disagreeable. This objection I have shown above to be ill founded, as gas is only dangerous when it is confined. The quantity *continually* discharged through the outlet would of course be infinitesimal. It is only by a system of sealed or closed traps and air-tight pipes as now used, and so persistently advocated by plumbers, that gas becomes dangerous, and in proportion to the free escape provided does it cease to be so. Our engineers too often insist on this air-tight condition of sewers and drains, but the evidence of the last half-century, and the increasing mischief of sewer emanations, show that this is wrong, and that our present glazed socketed earthenware and iron pipes and brick sewers, in which few vents exist, require safety vents, and these points of relief to our gas-charged systems are highly necessary.

Summing up the advantages of the "open junction" over the ordinary pipe ventilation, we have the following:—

1. *Greater efficiency.*—The escape is provided just where required, over and in a line with the soil-pipe, and before it enters the syphon-trap, and it provides an aperture of the same area as the pipe.

2. *Facility for cleaning the trap or pipe.*—By simply removing the perforated top or grating access is readily obtained to both trap and pipe, and any stoppage removed.

3. *Economy.*—It saves the expense of a lead or iron ventilating-pipe carried up to the roof. A ventilating-pipe, unless carried up a fire flue, frequently creates a cold descending current, particularly under certain states of the atmosphere, and then checks the escape of gas or forces it through water-trap.

The ingenuity of inventors has generally been directed in making gas-tight traps, as if the pressure of gas could be limited. The present invention secures the water-trap as a useful secondary check, at the same time affording that relief for the confined gas without which a thin layer of water is of no avail. Giving free escape to the insidious foe, it at the same time makes the water-trap of greater use and efficiency, for it should be known that water absorbs gas under pressure, and when saturated or impregnated with gas must evolve it. Cistern water has been poisoned in a like manner by the gases conveyed through waste-pipes. These are some of the manifest advantages afforded by the "trough ventilator." Giving to every house perfect immunity from the poisonous effluvia of the sewers, it also affords ready means of cleansing pipes.

We may caution the young architect against the employment of all complicated arrangements which the ingenuity of patentees has devised, and to content himself with the simpler contrivances which commend themselves to his common sense and good judgment. Many very excellent ventilating traps and systems of house drainage are before the public; we prefer those which are easily accessible and allow free escape to the gas. The germ of the best ventilating traps is an open syphon, and we may mention Buchan's ventilating drain trap as an effective appliance based upon this principle. It can be used to ventilate the sewer and to allow a current of fresh air to pass through the house drains. Messrs. Doulton supply stoneware traps for the same purpose, the object in all cases being the disconnection of the house drains from the sewer. All drain pipes should be true in bore and smooth, and a good fall is 1 in 60 or 2 in. in 10 ft., the inclination for the branches being somewhat more. Their size should seldom be less than 4 in. in internal diameter, but 6 in. is necessary in most cases, so that solid matters may find an easy passage. A 6-in. drain is sufficient for several houses when the rain-water area is not large.

For the admission of fresh air to buildings we recommend the conical-shaped perforations lately introduced by Mr. Ellison of Leeds, by which the air is arrested and diffused as it enters the apartment. Bricks and skirtings are manufactured on this principle. For outlets we have already mentioned the deflecting or exhaust ventilators, of which there are several in the market, applicable for every variety of purpose.

The excellent closets of Mr. Jennings, Doulton, and some others, are recommended for simplicity of construction, cleanliness in action, and freedom from the liability to corrosion. They are made of one piece of earthenware, and have in most cases a syphon-trap combined with the pan. In all instances the old-fashioned pan closet is to be avoided; the plug valve and direct pull being preferable. In the case of waste-preventers, simplicity of action should govern the choice, and many excellent patents are to be obtained.

Since the foregoing observations were written, I have had several opportunities of examining the corrosive action of sewer gas on lead; in some instances, the traps (especially the old D trap) were found honey-combed, or perforated, at the upper part by the action of the gas.

These observations lead to the conclusion that—

1st. Lead is not altogether so desirable a material as plumbers would have us believe. Plumbers prefer lead because it is the material they are most acquainted with; and because it admits of easy jointing, and is more pliable than iron, &c. Some of our sanitary manufacturers are, however, wisely giving preference for earthenware and tin-encased lead pipes for traps, and the pipe and drains for the conveyance of soil, &c. Glazed earthenware closet traps are decidedly cleaner than lead, and not liable to corrode. Closet pans and syphons are now made in one piece of ware. (See Jenning's patent.)

2nd. We are led to the conclusion that any trap or pipe which harbours the gas, or encloses any space or corner where it can collect, is defective; hence D traps, when unventilated, are more liable than S or syphon traps, to be corroded or eaten through by the gas.

CLOSET APPLIANCES.—It may not be out of place here to speak of the closet appliances in use. There is the "pan closet," so called from its having a large pan and container under the basin. It has disadvantages from which its rival, the "valve closet" is free. The container or receiver below the basin is apt to become foul, and quickly corrode, and it is further open to the objection that it collects gas in its upper part. In cases where it is used a small ventilating-pipe may be inserted, communicating with the soil-pipe ventilator or shaft. To remedy this liability to become foul, Messrs. Warner, of Cripplegate, London, have provided china containers in lieu of cast iron, which so soon rusts and becomes coated with offensive matter.

The "valve closet," having no enclosing pan, is more direct and clean in its action, the soil at once entering the trap, and the flushing is more concentrated and perfect. There is, also, no space for the accumulation of foul air as in the "pan." India-rubber-faced valves are sometimes fitted to these closets, making the apparatus noiseless—a great desideratum. The overflow-pipe usually provided to basins of this class is often an injurious addition better avoided, as, unless ventilation be provided as recommended, it only conveys effluvia into the house. The Hopper closet, consisting of a simple basin and trap, or Jennings' earthenware closet is the best.

The overflow-pipes of cisterns are a fruitful source of mischief. They often are untrapped, and connect with unventilated soil-pipes. A small bell-trap is often used, which is of no value whatever, as it soon gets empty by evaporation. A self-acting trap is essential in these cases, especially where the cistern supplies drinking-water, though a separate cistern is best for the latter purpose.

SANITARY CONDITIONS. — We may sum up the necessary conditions of house construction as follows:—

1. *Dry subsoil.* — Houses should not be placed

against the foot of a slope, especially where a bed of clay crops out. Basements should be cut off from the surrounding ground by a bed of concrete or asphalte, or dry areas below level of basement floor, so as to ensure an open air space all round. Basement floors should be hollow, well ventilated over damp soils. Damp-proof courses should be provided in walls. Drains should never be under a house, but should be accessible; if under, they should be covered with concrete to make them air-tight. Sinks and wash-houses of cottages should be outside walls. Waste-pipes should empty into trapped gratings, so as to leave air-space between outlet and trap.

2. *Pure air.*—All drain and soil pipes should be trapped and ventilated. Cesspools should, in all cases, be ventilated separately, and formed in the lower surface of strata to prevent percolation into live wells.

3. *Warmth.*—Walls and roofs should be of hollow construction, and non-conducting materials, and be provided with air gratings. Floors should be rendered fireproof and sound-proof by concrete filling-in between joists or by pugging. Windows in exposed sides should be double; entrances should be protected by double doors.

4. *Ventilation, &c.*—Ventilation should be provided in all rooms by a proper provision of inlets and outlets so as to circulate the air without draught; attics, and apartments in roofs, and staircase lanterns should particularly be ventilated. Trap-doors should be provided to roofs.

Cisterns should be covered by felt, also all pipes that are external, with means of access for repairs. Hall's water-waste cistern avoids liability to frost. All drains at junctions and bends should have movable lids.

5. Lofty buildings should be protected by lightning-conductors; all stoves and ironwork being connected by wire to a well in the earth, or to the water and gas mains.

PART VI.

DESIGN.

SECTION I.—ARCHITECTURAL TASTE.

BEFORE proceeding to consider the principles upon which Architectural Design is based, it may be as well to make some remarks upon the general tendency of art-thought and the transition of taste which the last century has witnessed. These changes or "fashions" cannot be too thoughtfully considered by the young architect, as indicating both the natural tendency, or reactionary impulse, of the mind for variety; and also as affording a very reliable means of testing the fictitious, and discriminating between truth and error; between mere sentiment and the immutable laws of design in art. Since the publication of the first editions of these "Hints" architecture has undergone considerable changes in regard to its manner or style. Greek art, and the traditions of classic Italy, were at one time the "rage." Architects pinned their faith to the "five Orders," and few ventured to depart from those proportions which the finest examples of Greece and Rome afforded.

Every column, moulding, and intercolumn was nicely proportioned by the "module" or diameter of the column to within a "minute," or sixtieth part; and the student who could accurately draw the "five orders," and define every detail, was considered to have acquired the rudiments of his art, which were equally applicable to churches and theatres, the nobleman's mansion, to the interior finishing of a room, or a garden temple. Stuart and Revett's "Athens," Vitruvius, and Palladio, were accounted indisputable authorities, and their dicta were followed and laid down with a precision that now appears ridiculous. A just admiration for the beauty and proportions of the columnar structures of Athens—the Propylæa, the Parthenon, the Erechtheum, and others; the consummate art and intellectual refinement shown in the subdivisions of architrave, frieze, and cornice, and the finer gradations of the entablature—the triglyphs, mutules, &c.; the entasis, or swell of the column, and the other niceties of adjustment in the forms and mouldings which learned and painstaking explorers have discovered; all these excellences, no doubt, fired the enthusiasm of the architect, and were sought after and reproduced with an untiring zeal by a generation of artists who have left an imperishable name.

While we are compelled to admire the matchless proportions and arrested lines of Grecian architecture, we cannot altogether overlook the error of reproducing with mechanical correctness the features and peculiarities of a style which were not the most suitable to our climate and wants. An untiring devotion and enthusiasm for this exquisite refinement of art, and a keen perception of the merits of the style, led to a closer adherence to precedent than was consistent with our wants and correct judgment; and hence a reaction of a different kind followed. Roman models became fashionable; a less rigid style succeeded, and eventually gave place to the Gothic Revival.

While correctness and method were observed almost

to a fault, and the dogmas of classic art and its forms repeated *ad nauseam*, a change began which entirely set aside rule and precedent for a mode of thought and expression in which a picturesque irregularity and a sentimental disregard for order and critical perception were the predominant characteristics. Both in literature and art the change was marked. A love of the marvellous and chivalrous, the associations of a romantic life, incited by the writings of Sir Walter Scott and Byron, displayed the general tendency of the popular taste.

Again, in art, rectilinear severity gives place to curved forms, and graduation to picturesque fancies and broken masses. We find a complete transformation both in thought and art expression. The unbending dialect of Johnson is succeeded by a vernacular idiom. The Greek or Roman temple is replaced by the mediæval abbey or cathedral. Buttressed walls take the place of porticoes and colonnades, pointed arches of square lintels; and the fritter of meaningless adjuncts, pinnacles, and crocketted spires, exchange the rigidity and rectangularity of unbroken sky-lines.

A frivolous seeking after effect became as painfully apparent as the monotony of columnar façades, and the concealment of roofs and gutters by false parapets, pediments, and balustrades.

From the toy-like Gothic of Walpole's time,* with its gabled, castellated, or pinnacled features, the conventional type of church, with its attenuated windows and western embattled tower, of which there are many examples still existing, to the later type of fashion in which the continental Gothic is predominant—a noticeable change of character and a considerable leap is manifest.

* Known as "Strawberry Hill" Gothic, introduced by Walpole there in 1770.

From overwrought and ornate examples, based on the Tudor, or Late Perpendicular, we were lately deluged with the revival of foreign varieties of Gothic; crude, meagre, and frequently devoid of any of the higher and more intellectual beauties and refinements of Gothic.

After ransacking English Gothic examples, and studying Pugin's, Rickman's, and Parker's works, hurried and thought-grudging architects have left our own shores for examples and types less pure, or the contemptible motive of mimicking Italianisms or French forms of Gothicesque.

The student is earnestly invited to consider thoughtfully these changes in the architecture of our era; to discard the poetic romanticism and extravagances of the art, even though under the sanction of our most fashionable pseudo-Mediævalists; to weigh well the merits of those several schools of art-thought which the last three centuries have revived; to consider what elements in them are worth adopting, irrespective of narrow prejudice and sectarian ideas; and so to evolve, from a generalised study of the past, that particular mode of construction and expression which best fulfils the requirements of the present age. Let the word "style" be banished from his mind—his practice at least—so that he may *follow out* his thoughts without a conventional rendering or phraseology borrowed from a bygone age. Let him discard the traditions and dogmas of Mediævalism as regards the arrangements, forms, and accessories of his churches; and, trusting only to his common sense of fitness, adapt his materials without reference to mere type, restrained only by a rational use of the fittest material and the most scientific method of construction. We do not mean that novelty should usurp experience, or be sought when a certain type or precedent can be found equally adapted to our wants. This would be rashness and presumption.

Let the student particularly avoid *affectation* or mannerism as the very curse of Art. Every true artist must henceforth be a school in himself. The styles of the Greeks and Romans, the Romanesque and Gothics of the West, the styles of the East, of the Renaissance, or the schools of the Revival, have all served their time and departed. If we recall them, we can only affect their manner without adding to our art one idea. The Periclean and Augustan ages, the thirteenth and fourteenth centuries, with their wondrous arts, cannot be brought back; we may as reasonably expect to roll back the tide, or recall the past. These great phases of architectural thought were the results of working priesthoods or classes—the *thinking*, simple, honest-minded mason, craftsman, or artist. There was not the art-cant and thoughtless hurry and affectation we have now. Our modern taste is vastly different. We copy because we will not think for ourselves; our national architecture is a reflex of our multifarious ideas and tastes, rather than of our wants. Unlike the arts under Pericles and our Edwards, ours has no basis or belief to repose upon; it belongs to a nation, not a class, or society, or dominant Church; it has all the elements of discord, it adopts anachronisms and styles of the most opposite and incongruous description; chiefly, from a want of agreement between the material, mental, and moral states of our civilisation.

Instead of resting on a subjective basis, modern invention, as respects art, is utterly feeble, and has failed to realise that balance of the intellect and feeling which alone can produce the true and beautiful. *Thought*, then, should preside, and influence all our architecture; it should be freely not grudgingly bestowed upon the smallest detail, no less than upon the general design of a structure. We should take nothing without this crucial test—nothing without its sanction. True Art is the combination of the *constructive* and *æsthetic*—intellect and moral power; its object must be attained in the *simplest* and most unaffected manner, not by following, as our pseudo-Classicists and Mediævalists have done, any past style, nor by mistaking barbarism and crudity for simplicity, as our ultra-Revivalists are now doing. Whatever may be the principles we take for our guidance, we may be sure nothing will atone for the want of thought in our designs. Thought-sparing art has been the source of our decline in constructive and decorative truth. We are flooded with Continental works of art; our students have been taught to copy and follow instead of work and think for themselves. Italian and French Gothicisms are thrust upon the young architectural aspirant before he can master the materials and principles of his art; he is apt to follow this or that leader of "fashion," as his imagination tends, and thus his future career is fettered and cramped by a false system of conventionalism and slavery, instead of prompted and dictated by a free and independent power of thought, the result of a catholic and nobler view of art, and a generalization of nature.

Let the student of this art regard the mutations which have successively marked its history during the last century—the Grecisms, Italicisms, of the Adamses, Soane, Nash, Wilkins, Barry, and a host of departed classicists, the Gothics of Walpole and Pugin, and our own ultra-Mediævalists—the ages of pseudo-temples and lath-and-plaster pagodas, as merely passing and reactive changes, resting his taste on the broader principles which underlie all true art—the primitive impulses that create it.

Section II.—Principles of Design.

Every perfect work of art should satisfy and correspond to the necessities of our nature, namely, the Physical, Intellectual, and Emotional states.* The two first of these may be considered as embraced under Construction, and include the principles of *Utility* and *Fitness*, which lie at the root of all good design. The emotional or æsthetic spring from or grow out of these.

In the present brief section we will lay down some of those general laws which should determine the design of various functions of architecture, and also the legitimate employment and use of materials.

The multifarious uses and purposes to which Architectural Design is applied are not now considered, as they belong to a wider view of the subject than can be discussed here, and further imply questions of individual and special requirement, discipline, and the ever-changing and progressive notions which commerce, science, hygiene, and improvements in mechanical appliances and materials call into existence; as in the arrangements of our domestic and commercial buildings, educational institutions, structures devoted to public worship, as churches and chapels, and the various buildings erected for sanitary purposes, hospitals, asylums, unions, prisons, and the like. All these kinds of structure, it is evident, demand special wants, which only the experience gained from an examination and study of actual works of merit of their several kinds can supply. They all require a knowledge of the best arrangements and other concrete kinds of information with which the young architect should make himself familiar by the study of works specially

* See a work entitled "Theory of the Arts," which first appeared in the *Building News*.

devoted to their consideration, and the statistics and reports of parliamentary evidence contained in various Blue-books. Our duty now will be confined to those abstract principles of architecture proper, which deal with structural and technical facts and requirements.

Different conditions and circumstances demand solution in every case in which an architect is consulted. Thus, the extent of land, its shape and surroundings, special wants, as entrances, and modes of lighting, contiguity of buildings, accommodation, materials of neighbourhood, and not least, cost, must in every case determine the arrangement, position, and construction of a building, in all of which cases it would be impossible to lay down any general rules.

Design is not dependent on abstract form merely, such as implied in those theories which are founded on plant form, geometrical ratios or proportion, harmonic laws, &c., which all err in not regarding the *concrete* nature of architecture.

The very complexity of conditions the architect has to deal with, is the main cause of his neglecting, as too often he does, the dictates of direct science. The deductions of science required by the engineer can be *directly* applied, immediate utility being the only end sought; hence he works by more exact laws, and the results are therefore more satisfactory and determinate. On the contrary, the architect, besides utility, has to gratify the emotions, and thus he too often works more by his imagination than his reason, more by custom and precedent than science. His art stands in much the same position as that of medicine, half empirical, half exact, and the laws of his art are in the same undefined category as those of moral science, occupying a middle ground between the exact and indefinite. Thus he is continually waging a war between advanced theories and the demands of science and those of conventional art—between the progressive and unprogressive elements of his art.

LAWS OF DESIGN.—The laws which should govern architectural design are—

1st. PHYSICAL LAWS in which the senses of *weight, strength, security*, and *comfort* are concerned.

2nd. INTELLECTUAL or DISCRIMINATIVE LAWS. These control and adapt the materials and construction of a building so as to present to the mind a sense of fitness and adaptation of means to ends. Professor Bain, in his able work on "The Senses and Intellect," shows that our feelings for "differences and agreements" constitute the functions necessary in the performance of the intellect, and the discrimination has been expressed as the law of "relativity." Thus it is our discriminating perception of weight and support, which we call just proportion, that enables us to give to a structure an appearance of *fitness*, the mind being satisfied with the result. Such is the justness of proportion between the entablature and columns, or between arches and abutments, height or mass of wall and its supports, lengths and widths of apartments, &c. The adaptation of a material to its use is another example of this faculty of design.

MECHANICAL PRINCIPLES OF DESIGN.—All architectural construction involves ideas of *rest* and *motion*, or static and dynamic conditions direct and indirect support. In the vertical wall or pier we have the first typified; in the arch, dome, roof, and vault the second typified. The lintel, although usually placed in the first or static condition, is really more significant of the latter.* The Egyptian and Greek architecture

---

* Some writers, incorrectly I think, confound the abstract expression of lines and masses with their relative duties. It may be observed,

are eminently symbolic of the static principle; while the Roman, Romanesque, the Gothic, and all arched styles, of the dynamic. Correctly, we think the latter styles and all true architecture show equally both principles. Now it may be noticed that the *static* principle was the earliest developed, as it was founded upon the law of gravity in its simplest form; it was in fact the simplest mechanical condition of building and the easiest lesson of experience, and one favourable, as we have shown elsewhere, to the subjective law of thought. It was extremely monumental, easily lent itself to this form of composition, as it did also to the decorative, for the conditions necessary to ornament were speedily attained.

To span a chasm, or cover a large area, with few points of support, required a new principle of construction, an economical balance of counteractive forces. The arch was the simplest and earliest expedient which answered this purpose. It enabled small stones to be used, and we find the columnar and monolithic construction gradually giving place to a system in which the arch and buttress became the leading elements. We have here the static and dynamic principles combined, each co-operating with the other. The constructions of the Romans first developed this system, which was perfected during the Middle Ages, and our own modern works have carried the arched system to a very advanced state.

The highest forms of architecture—the Romanesque and Gothic—have therefore both the *static* and *dynamic* conditions in a state of balance and agreement; the columnar and vertical masses (static) being everywhere in conjunction or acting together with the active

that horizontality can only imply the idea of rest when connected with *vertical* support, and that a long unsupported beam is anything but stable.

weight or thrust principle of the arch and vault (dynamic) —never, however, in a state bordering on a tottering equilibrium, but so far removed from it as to evince mutual counteraction and aid; thus we have the buttress sustaining the nave or aisle vaulting, and yet aiding in reality and emphasizing in effect the static repose of the structure or the vertical parts. The graduated or staged buttress (which modern Gothicists are surrendering in favour of the pilaster or Italian buttress) exemplifies this principle. While it performs the office of abutment or counterfort to an unseen but still-evident force created by the arched opening, vault, or roof (the graduated increase of thickness as it reaches the ground agreeing with the theoretical "line of pressure"), it also affords economic support and expression. The pyramidal outline of Gothic buildings displays mutual and co-operative support of the two principles which we fail to observe in the simple static structures of Greece.*

As characteristic or symbolic of these principles it may be noted, that while the static or columnar forms are simply *rectangular*, as in the Greek temples, the dynamic is evidenced in *triangular* and curvilinear forms. The economical adjustment of material and forces has thus, in all true composition, shown or expressed itself in the leading and dominating lines of features and mass, and indirectly also influenced and given form to every detail. The mechanical framework or contrivance has also been the chief impulse to decoration, and has controlled imagination. Under our present system or fashion of design, the forms are repeated without the correspondence of this initial or motive impulse. The following conclusions may be deduced:

* It may be observed that in Grecian architecture the predominant lines were horizontal and continuous, and the subordinate lines vertical and discontinuous.

1st. The highest forms of architecture are those in which the *static principle* (or repose) predominates with the *least* amount of material, or in which the static and dynamic forces and expressions are pleasingly balanced, as in the Greek and Gothic.

2nd. The correspondence of the mechanical and decorative, or the intellectual and imaginative, elements is necessary.

The Egyptian and other primitive structures fail in the first: the material preponderates. In the Parthenon the static principle is more evident to the senses than the mere mass of the pyramids or the Hall of Karnak, which expresses only weight and immovableness. In the mediæval cathedral, as our Salisbury, Lincoln, or Amiens, we have the balance still more delicately expressed. The Indian and some other forms show the dynamic condition only.

MATERIALS AND THEIR FUNCTIONS.—We have three distinct structural elements, which play important parts in architecture:—1st, the *column;* 2nd, the *beam* or *arch;* 3rd, the *tie*. Each of these has been developed under different systems of art. The first we find in the columnar expressions of Egyptian, Assyrian, and Greek examples combined with the beam. The second, or the arch, we have in the Roman, Romanesque, Byzantine Moresque, and Gothic developments; while the third, the principle of the tie, has been reserved for the latest necessities of architecture, and exemplified in the *Truss*.

It is evident there are certain materials that lend themselves to these special elements. For example, *stone, brick*, and all homogeneous granular substances are peculiarly adapted to compressive action, as in columns and arches; other tenacious substances, such as timber, iron, and metals, adapt themselves to tensile action more favourably than the former materials. The beam, however, is subject to two separate actions or strains—a compressive, in the upper section, and a tensile in the lower part. This double action makes it necessary that a beam should be of a material which can equally resist both these strains. On this account, a stone beam, if too long in proportion to its depth, is out of place, and weak; it would be liable to open below from want of cohesive resistance, the resistance of stone or brick to tension being only about a fifth of its resistance to compression. For lintels and very short bearings, as exemplified in the classic entablature, stone may be used; but its legitimate purpose as a material is for walls, piers, and arches, and for columns whose height does not exceed ten or twelve times the diameter. Timber and iron are the materials which combine the compressive and tensile resistances to the greatest extent, and therefore are chiefly adapted for beams and other uses, in which these two kinds of strain are called into exercise. Compound beams of brick and iron or concrete and iron are sometimes used with advantage, as in flooring, roofing, &c., but in these cases the materials are combined so that each may have its proper function. From these considerations, then, it appears that all *stone* and *brick work* should be designed in conformity to the following principles:—

1st. They should be placed in positions, and perform offices, in which only weight or compressive strains are concerned, as in walls, piers, columns, arches, &c.

2nd. In designing stonework or brickwork, *mass* and *weight* are required, and forms should be given which best meet compressive action. Tensional stress should be avoided. Thus in buttresses we require depth as well as thickness. Piers and columns of stone or brick are best of rectangular, square, octagonal, circular, or other figure of section whose geometric centre is symmetrical, or nearly, and so designed as to produce at every bed-joint uniform pressure. The piers and clustered shafts of Romanesque and Early Pointed work illustrate this principle. The arch moulds, though often complicated, are in good examples designed in section within rectangular outlines, as are also the engaged and detached shafts which follow the simple square outline of Norman recessed jambs and archivolts. The plans of mullions also illustrate this principle; in the early and better examples, the sections are nearly square with the angles taken off, or boldly beaded or shafted, while in late and debased work hollows, fine fillets, and deep quirks destroy the real stability and apparent strength.

Where the pressure is one of varying intensity, as in the case of a pier supporting a thrust, this should be considered, and sufficient projection and mass given to that part of work upon which the greatest pressure is thrown. In the case of an abutment or jamb, against which an arch abuts, *visible* as well as actual security should be given, or the apparent effect will be to make the pier weak, if not bent to the eye; flat segment arches springing from narrow piers always give a crooked appearance to the piers, although architects seldom seem to notice or heed it.

3rd. Acute angles or arrises should be avoided. No angle should be less than a right angle in constructive masonry or brickwork, and sharp angles are inconsistent with the nature of such materials, even when used decoratively.*

IRON and TIMBER, when used to resist compressive forces, admit of somewhat wider limitations, being

---

* For further details the student is referred to the excellent treatise by Dobson, "Masonry and Stonecutting." (Lockwood & Co.)

less friable in their nature. Moulded forms may be employed more consistently. For tensile strains sections of less area are allowable, though in all these cases, square, round, or polygonal figures are more suitable. In the section on "Construction" we have indicated the best forms for beams, columns, and girders. The following principles should be observed in designing ironwork :—

1. Columns, struts, and other constructive ironwork subject to compression, should be of cast iron, of uniform or symmetrical section, either solid or hollow, round, cross-shape, square, or ┬-shape section.

2. Beams subject to cross strain should be designed with special reference to the material. Cast iron, having a greater resistance to compressive than tensile strains, should be designed with larger sectional areas or flanges at those parts subject to the latter. Inequalities of thickness in the metal, and sudden changes of form, should be avoided.

3. For work subject to sudden jars, changes of temperature, cross-strains, and where tensile strength is chiefly required, wrought iron should be used. The same sections may be adopted for beams, having regard, however, to the ratio of tensile to compressive strength. Lightness rather than mass should be observed in designing all wrought-iron work.

For WOODWORK the following rules should be observed :—

1. Columns and all pieces subject to pressure should be of square or rectangular section, plain, chamfered, or simply moulded. Circular and octagonal sections may also be employed, but are not so suitable or characteristic of the material. For example, in framework, turned pillars, with caps and bases of wood, are not so rational a treatment as chamfered or moulded

work, as such forms do not lend themselves to the natural fibre and grain of the material.

2. The *ornamental* treatment of woodwork should invariably be dictated by its fibrous nature. Cut, perforated, or turned work should only be applied to ornamental purposes, and on a small scale. Deeply indented or cut timber, in which the grain of the material is weakened or impaired, is improper; deep quirks, or deeply cut members, should be avoided. Mouldings, chamfers, and bevels are admirable for decorative purposes, while deep quirks and undercuttings are unsuitable as harbouring dust and dirt, and for external woodwork are injurious on account of retaining moisture, and tending to accelerate decay.

For carving and turning, the denser kinds of wood, as oak, are more suitable.

PLASTIC SUBSTANCES, such as cement, plaster, and other compounds of a tenacious kind, *carton pierre*, *papier-mâché*, &c., should be treated in the manners following:—

1. Casting, as in the cast or moulded enrichments and accessories of building.

2. Running, as in moulded work, cornices, strings, &c.

3. Stamping or impressing, as in ornamental panels and shallow surface relief.

Cast work should never be too deeply undercut. *Bas-relief* is more adapted to these materials and the mode of manufacture than alto-relievo. On the whole, we think running or moulding and stamped work the most characteristic treatment for cements and plasters, as high or deeply cut relief is evidently inconsistent with the brittle nature of such substances.

Ceramic materials, as terra-cotta, earthenware, porcelain, &c., admit of higher relief for ornamental purposes, being employed in small pieces as panels, string-courses, &c.

CAST AND STAMPED METAL WORK, as zinc and other alloys, should be designed as indicative or expressive of such modes of manufacture, and not made to represent carvings and chisel sculpture.

In short, every material, natural or artificial, should firstly, be adapted to its proper function and purpose; and secondly, the artistic or decorative treatment it receives should express rather than conceal or counterfeit its nature.

The tirade against shams of " stucco " and " compos," which devices prevailed during the early part of this century, has called forth a more truthful exhibition of material; yet it must be confessed we employ our materials often more regardless of structural conditions, as, for instance, when we use soft brick for label mouldings and other exposed parts in a stone building, or condemn cement and plaster where they would be of service.

Having discussed the functions of architecture, the concrete conditions of design, and laid down general principles which should govern all constructive members of buildings, we may now briefly recapitulate those *abstract* elements of design which have reference to Form only, and which please the eye, and through it the mind, by awakening the emotions. These may be classed as purely *æsthetic* elements, and occupy the border-ground between the Sensuous, and Intellectual, and Emotional. Such principles constitute all forms of beauty, and are—

   I. Proportion.
   II. Unity, or Symmetry.
   III. Variety.

IV. Contrast and Gradation.
V. Optical correction.
VI. Expression.
VII. Colour.

BEAUTY OF FORM.—Without discussing the various hypotheses of philosophers and writers on art as to what constitutes Beauty, we may say that it depends on a combination of properties which appeal to the higher emotions rather than upon any one peculiarity or element inherent in the material, as Burke propounds; on any associations of a pleasing kind, as attempted to be laid down in Alison's theory; or upon any distinct principle of an abstract kind, as a line of double *flexure*, arbitrarily defined by Hogarth and others. All these able theorists, however, have advanced hypotheses more or less near the truth, and their works may be studied by all architects and others engaged in the arts of design. In our own day, Mr. Fergusson, in his "Principles of Beauty," who classifies all beauties under the three heads of (1) Technic, (2) Æsthetic, (3) Phonetic, thinks architecture (as the Greek) combines in equal proportions these three kinds of beauty or merit. As justly observed by another writer on the "Principles of Design,"* this categorical theory does not define the *kind* of merit, only the *degree*, and that we must look for a class of art possessing an *expression* peculiar to itself, not associative, as, for example, in confounding Gothic art with Religion, or Italian with the Secular, as commonly done. The great point is to distinguish the *accidental* and *local* from the *universal* and *essential*, or the generalised expression of merits which aptly find their

---

* See the able work on "Design" by E. L. Garbett. (Lockwood & Co.)

counterparts in our physical, intellectual, and emotional faculties.

Professor Bain, in his work on "The Senses and Intellect," and his "Manual of Mental and Moral Science," has shown clearly the dependence of our emotions upon organization, and we refer our readers to these works for a philosophical investigation of the causes of beauty.*

Briefly, the beauty of form depends upon an agreeable balance of those qualities which satisfy the mind and emotions. A work of art may be striking from its size, outline, or some sensible property it possesses, yet it may not be beautiful,—some other quality may be needed. If it have in addition certain qualities or attributes which combine to give it expression, grace, and fitness, and which may depend on various merits and sensible properties, we then call it beautiful, or perfect. It is evident every one cannot realise the same standard of beauty, though there are some abstract properties which are universally regarded as beautiful. Some see beauty in the Greek Parthenon, others in the Mediæval cathedral. We must therefore look for other elements which are necessary to the beautiful.

I. PROPORTION is one all-pervading influence. By proportion we mean the adjustment of weight and support, ratio of height to breadth, or fitness of size and bulk for the intended purpose.† Constructive fitness lies at the basis of it. Attempts have been made by ingenious theorists to apply certain arbitrary ratios to architecture, founded upon analogies of sounds, colours, &c. Thus Vitruvius gives a certain ratio for the height and breadth of a window, because two strings of a like

* See Herbert Spencer's works.
† A room whose length and breadth have some simple ratio, as 3 to 1 or 3 to 2, is more agreeable than when no ratio is discernible.

ratio will give harmonious notes. The analogy is far-fetched, yet talent has been misspent in attempts to lay down definite forms and proportions founded upon harmonic laws, botanical geometry, &c. Some mental cause must be found for all preference of one proportion over another.*

II. UNITY and SYMMETRY are important elements of beauty. Correspondence and connection of parts, equal spacing, &c., may be embraced within this category. *Oneness* of form, number, ratio, succession, are equivalent terms implying *method* or *order*. Dr. Hutcheson remarks as regards formal beauty, that when uniformity is equal, the beauty of forms is in proportion to their variety; and when their variety is equal, it is in proportion to their uniformity. A curved line of irregular flexure is not so beautiful as a regular one, as an arc of a circle whose direction though changing does so *uniformly*.

III. It is the combination of Unity with VARIETY which is needed, and this we get in the simple and varying curves, as the circle seen perspectively, the ellipse, hyperbola, parabola, catenary, spiral, &c. GRADATION is another term for this combination.

IV. CONTRAST may be defined as a change of resemblance, either as a change of shape, position, &c., but resemblances must exist as well as differences. Hogarth's line of beauty, being a curve of contrary flexure, is an apt example of the principle of variety, gradation, or contrast. It has been well remarked in the case of curves of opposite curvature, the best

* Mr. D. R. Hay has applied the Pythagorean system of harmonic numbers to architectural proportions, the basis of the theory being that a figure is pleasing to the eye in proportion as its primal angles bear to each other analogous proportions to those of vibrations in a chord of music, and that harmony of form arises from a simple division of a quadrant or right angle into harmonic parts, as $\frac{1}{2}, \frac{1}{3}, \frac{1}{5}, \frac{1}{7}$, and their multiples.

contrast is obtained by equality of curvature, or when the curves are both equally deflected from a common tangent. The most graceful Greek kinds of ovoid form exhibit this equality in the opposite curves.

Nature expresses force, strength, and exciting qualities by angularities and contrasts; the more delicate feelings and qualities by curvature and gradation (see Alison on "The Sublimity and Beauty of the Material World," ch. iv., part 2). One general and valuable principle may be deduced, viz.— all structural parts of a building or work of art should have the more forcible elements of expression, and the minor and ornamental portions the most graceful and elegant kinds of form.

Thus the perfect Greek and Gothic show a remarkable contrast; in pure Greek we have *interrupted vertical* lines, in Gothic *interrupted horizontal* ones. In all composition both vertical and horizontal connection is necessary to preserve unity of parts.

Other considerations should govern the kind of form best suited, such as the *destination* of the building, whether it be of a grave or light class, and the relative *position* and *importance* of the features. Some of the principal kinds of expression in abstract form arising from the principles of Contrast and Gradation are—

(1) Rectangular forms, or forms bounded by planes or lines at right angles (a right angle being the strongest contrasted angle).

(2) Forms bounded by oblique lines or planes, expressing an intermediate state between strength and force and delicacy or grace.

(3) Curvilinear forms more or less contrasted, as in geometrical or flowing traceries, forms of arches and roofs exemplifying gradation and ease, and

K

all the lighter and more soothing qualities of expression.

(4) Mixtilinear forms, in which the above expressions are blended, as in the combined bead and square or roll and concavity.

V. OPTICAL OR OCULAR REFINEMENT can hardly be called an æsthetic principle of abstract form, yet it plays an important part in design. Straight lines joining curves, as in some Tudor or flat-pointed arches, always look concave, and have an apparent deflection at their junction very disagreeable. Straight-sided shafts, as columns, spires, chimneys, &c., look weak and concave-sided, which is corrected by a slight swell, or convexity (entasis), which the Greeks applied to all the horizontal and vertical lines of their temples.

Again, a slight camber, or rise, to flat-headed openings, prevents the sunken appearance a perfectly straight lintel always has, and is necessary in all horizontal and level lines and surfaces.*

Height is obtained by the predominance of vertical lines, and lessened by horizontal lines. Apparent size or magnitude is obtained by multiplicity of details, or small patterns, forming a kind of scale to the eye, and *vice versâ*. Contrasts always tend to exaggerate the difference.†

Decided contrasts of form or direction are often necessary to correct ocular impressions by making differences appear greater.‡

The *irradiation* of light, or the effect of luminosity in spreading the lighter parts of objects, is an im-

---

* These effects are partly due to the spreading influence of light (irradiation).

† Careful measurements have shown that the axes of the corner columns of Greek temples slightly converged.

‡ Thus, unless exact centrality or uniformity of features be obtained it is more pleasing to increase the departure or difference.

portant principle in optics. On this principle the drops of the Greek tænia were made conical, and larger at the lower end, where they are seen against a bright surface. Various other instances may be noticed, but the reader is referred to other treatises, especially Mr. F. C. Penrose's able investigations on the optical refinements of the Parthenon.

VI. EXPRESSION.—Although placed here, this is one of the principal attributes of true art, but seldom heeded by architects unless mixed up with associations. The grave and festive, the majestic or playful, should be expressed by architecture to the simplest individual. The stern majesty of the hall of justice, the aspiring tendency of the religious temple, the festive place of amusement, and the urbanity of the domestic residence, should proclaim their respective purposes, at least in general terms.

VII. COLOUR.—This belongs to the last and lowest class of effect. Primitive nations have invariably, like children, been attracted by colour of the brightest hue. Forms, especially of a natural or sensuous kind, are next appreciated; and, lastly, those of a mental or conventional order. Monochrome decoration has consequently been only appreciated by a few; the positive colours of the polychromist being the most popular, because most pleasing and sensuous.

Children and savages have, as observed by eminent authorities, always shown a preference for positive colours; the nicer and less sensuous distinctions of shades and tones requiring greater mental discrimination.

Hence it may be laid down that nature offers the best rule for our guidance, namely, unity of colour, neutral tints, or secondaries and tertiaries for large and retiring surfaces and masses, and varied primitive colours in the smaller and prominent parts. To produce the effect of light and harmony, there should be a balance of colours according to the proportions of the solar spectrum. The eye exposed to one colour is most satisfied and pleased with its complementary; thus *red* harmonizes with *green*, *blue* with *orange*, *yellow* with *violet;* and these colours are harmonic, and are shown by physicists to have a *simple* ratio of vibrations to one another as between two musical notes. This ratio is 4 to 5 or 5 to 4. According to this law, the primaries should be employed in a certain proportion to one another, *e. g.* 3 yellow, 5 red, and 8 blue, the latter as the retiring colour being used in the concave surfaces, yellow which advances on the convex, and red as the middle-distance colour.

Classification of beauties.—All beauties or merits are resolvable into *ocular* or sensuous impressions, as the lower sort of elements, and the *intellectual* or *ideal* class. Many philosophers have denied that there exists any one *form* more pleasing than another in the abstract, *i.e.* inherently, intrinsically, and that any preference is simply the result of mutual inference and association with certain emotions. We have shown, however, certain elements or pleasing combinations to exist, though there is always the difficulty of entirely setting aside the idea of their association with certain objects; in other words, there are certain properties which affect the senses and mind independently of their awakening any extraneous emotion, *e.g.* a curved line is always more pleasing than a straight line, and a circle than a square.

The young artist will perceive that the higher kinds of expression spring from those qualities which are the result of discrimination and cultivated taste.

ADDENDA ON CONSTRUCTION.

MORTAR AND CEMENT.—LIMES are either rich or hydraulic, according as the limestones are of carbonate of lime, such as chalk, or contain from 15 to 30 per cent. of silicates or magnesia. Rich limes slake freely, and augment largely in bulk. They harden slowly in air, but not at all in water. Mortar made of them is soon affected by damp, &c. Hydraulic lime slakes slowly, and hardens under water; hence is adapted for damp situations. Some stones produce cements which do not slake, but when ground and mixed with water, will set in air or water almost immediately. Such stones contain from 40 to 60 per cent. of silicates. The hardening or setting of mortar is due to the gradual absorption of carbonic acid from the atmosphere, thus forming a crystallized carbonate of lime. In hydraulic limes the induration is effected, probably, from a union of the lime with silica and alumina, forming an insoluble crystallized double silicate. For concrete and thick walls, and damp places, through which air cannot permeate, hydraulic limes should be used. The blue lias is one of the most hydraulic and strongest of limes. Sand for mortar should be sharp, clean, and gritty; and not too fine. Good pit sand, or clean road grit, is best. Salt sand should be avoided Water for mixing should be free from organic matter, and used sparingly. For ordinary mortar for brickwork the proportions should be, 1 of slaked lime, 1 sand, and 1 smithy ashes. For rubble masonry: 1 slaked lime, 2 parts sand, and one-third part smith's ashes. The admixture of ashes facilitates absorption, and makes mortar stronger. Mortar should be used fresh. Portland Cement is generally used where strength is required. It is made of clayey mud pounded under water, dried, and burnt. It is three times stronger than Roman cement, and improves by age. The patent "Selenitic" cement can be used as a substitute for mortar and cement. It saves lime, and sets rapidly. It is a species of cement-mortar, and is an approved method of using prepared quicklime.

BRICKS, STONE, WALLS, AND ARCHES.—Vitreous and glazed facing materials should be avoided in ordinary walls. They conduct heat rapidly, and condense moisture to an unpleasant extent. Hollow walls, tied with cramps, obviate this, and allow of the use of non-absorbent vitreous facings to a larger extent. Hollow bricks, or cellular walls, are better than solid ones. Hollow bricks, or stoneware, make capital arches and vaults, light and non-conductive. Floors may also be constructed of such materials rebated or joggled together, supported upon encased rolled iron joists—the encasing being of the plaster, concrete, or some non-conducting material, moulded externally. The author has used this method with success. (See Guillaume's "Economic Building Brick.")

STONE.—Stones should be selected with caution, and be placed upon their natural quarry beds when used. Laminated stones should not be used in work exposed to weather, or in highly relieved mouldings. Gothic mouldings deeply cut are most liable to decay from this cause. Delicate mouldings should be executed in stones whose grain is favourable to them. Coarse sandstones are often unadapted for highly moulded work. Long stones set on end as in mullions, architraves, &c., are to be avoided as false construction.

The greater the number of beds the better, and all good Middle Age masonry conforms to this rule. When large or deep stones are used the bed-joints should be proportionately thick. This is strangely ignored by architects and builders. Stones expand and contract by heat and cold, and in large masses, or long lengths, as in columns, copings, steps, &c., frequent joints should be allowed, or they will fracture, or be forced out of line. A combination of ashlar and rubble makes bad walling; concrete and ashlar, or brick facings, are to be preferred if proper bond courses are used.

COMBINATION OF MATERIALS.—The different conductivities and unequal rates of expansion and contraction of different substances, caused by changes of temperature, oxidation, &c., are a serious cause of disrupture, fracture, and derangement in building, which architects seldom heed. Care should be taken to combine and adjust these materials, so as to prevent this cause of failure. Modes of joining, free space at bearings, and means of adjustment should be considered. Thus metals expand more than stone, and all girders and iron framework should have free play at bearings. Sockets should be provided for their ends, instead of being allowed to impair the stability of the supporting walls by entering and being fixed rigidly to them. Columns and girders of iron should also be encased in non-conducting materials, such as porous plaster containing animal charcoal or earthenware. Fillets of wood banded round with iron, perforated plating or netting, covered with plaster, will prevent the destructive action of intense heat or fire, and render the materials non-conducting and less subject to the inconvenience of condensed moisture.

Mr. Hornblower's patent fire-proof flooring is formed on this principle, and is to be highly recommended.

The stability of masonry and brickwork should depend mainly upon the pressure of weight and static equilibrium. Iron employed as dowels, cramps, &c., frequently does more harm than good in corroding or expanding. Homogeneous materials and construction are the best; and when different materials are combined, they should be made to act and react upon each other without destroying the structure in which they are used. Good conductors should be encased in bad ones with air spaces between; and movable bearings should be used in all long lengths of materials.

BOND.—In bonding walls, hoop-iron, or courses of bricks in cement, are preferable to wood bond.

CONCRETE WALLS.—Walls of concrete, either built in blocks or by the filling-in process, as by Henley's patent, are cheaper and more impervious to moisture, vermin, &c., than brick walls. They may be relieved by panelling, impressed ornament, and faced with stucco or cement. The compressive resistance of concrete may be usefully employed in combination with iron tie rods for floors and beams, as in Hyatt's system. The statistics of Captain Shaw, of the Metropolitan Fire Brigade, show concrete to be one of the best and most fire-resisting materials.

ROOF COVERINGS.—Tiles corrugated or ribbed make one of the best roof coverings. The Broomhall tiles are recommended for appearance. Slate is more conductive of heat. All roofs should be either felted or pugged. In large open roofs, as those of churches, the pugging may be filled in between the lathing of inside plastering or the boarding and the slate battens, by placing diagonal strips or battens over the laths two or three inches in thickness, and then nailing the slate battens horizontally upon those, thus leaving a space for the pugging. A good and cheap roof can be formed in this way. Plastered ceilings and roofs are better than boarded ones, being less resonant of sound and more fire-resisting. An open air space above collared roofs and ceilings is desirable for ventilation, and also for preventing the passage of heat and retaining warmth in the building.

CISTERNS.—All drinking-water cisterns should be placed in readily accessible positions for cleansing (not over water-closets), and be covered with felt to prevent effects of frost. Overflow and waste pipes should not lead into soil-pipes, but discharge into stacks or externally. All water-pipes should be cased in felt or sawdust, and be easily got at by pocket pieces or doors in wood casings.

WATER-CLOSETS, PIPES, &c.—The common glazed earthenware syphon closet-pan is the cheapest and best; no overflow should be allowed into valve-chamber or soil-pipe, as in some old apparatuses, which allow the sewer air to escape. Lead-encased pipes (Haines' patent) are better than lead pipes; and wrought-iron pipes with screw joints are next best. All water-closets should be placed in external positions or projections, and have an intermediate or isolating lobby for ventilation, &c. Stack pipes may be used as ventilating-tubes to soil-pipes, &c.

BASEMENT FLOORS.—Sleeper walls of earthenware are better than brick for joists, as it checks the rise of damp from ground.

DRAINS.—Invert of main sewer should be at least nine inches or a foot below that of house or branch drain. Doulton's patent invert blocks are to be recommended. Junction blocks of stoneware should be inserted in main drains or sewers to prevent bungling jointing by incompetent hands, and caps, or movable lids, should be provided to house drains at every bend to facilitate cleaning the drains. Pipes should be jointed with clay, or tarred gaskin inserted in the sockets. Earthenware soil-pipes built into the wall may be used, but not zinc. Connection between sinks and house drains should be cut off by allowing the waste to discharge upon an open outside gully. All drains should be ventilated at cesspools or dead walls, and disconnected on both sides of the building by proper ventilating gullies (see Sanitary Construction, Part V.).

PART VII.
MODEL SPECIFICATION.

INDEX.

		No.
General Clauses.	Notice to local authorities. Payment of fees	1
	Protective inclosures. Restoration of pavements	1
	Removal and shoring. Old materials	2
	Clearing, excavating, re-filling	3
	Drainage of site	4
	Underpinning	5
Foundations, &c.	Trenches for footings	6
	Artificial levels for pavements. Ground-making round building	7
Well.	Well-digging and building	8
Artificial foundations.	Concrete foundations	9
	Piling and planking	10

BRICKWORK.

Indents.	Indents in old work, and making good	11
General clauses.	Bond—quality of bricks—mortar—rise of courses —flushing — grouting—footings — outside and inside work—hollow walls—facing and pointing	12
Damp-proof course.	Slate or asphalte damp course	13
Arches.	Gauged arches to windows, &c.	14
Rough Arches.	Rough arches—relieving arches	15
	Inverted arches	16
Brick-facing.	Common brick facing	17
Brick strings, cuttings,&c.	Brick strings—fascias—cornices—cuttings, &c.	18
Fire-places, flues, &c.	Fire-openings—arches—trimmer ditto—flues, stacks, &c.	19

		No.
Fire-places.	Chimney bars	20
	Fixing grates	22
Coping, &c.	Tile—or brick on edge	21
Bedding timber, stone, &c.	Bedding-pointing—backing timber—stone-work—frames, &c.	22
Dwarf walls, &c.	Dwarf walls—piers	23
Brick-nogging.	Brick-nog partitions	24
Rough brick paving, &c.	Paving—flat—on edge—asphalte paving	25
Tile paving.	Encaustic tile paving	26
Clinker paving.	Clinker paving to stables	27
Vaults.	Vaults of brick or masonry—filling in, &c.	28
	Groined ditto	29
	Corbelled skew-backs	30
	Concrete filling in of vaults	31
	Cemented outside	32
Lime-whiting		33
Hoop-iron bond		34
Extra brickwork		35
Party wall		36
Drainage.	Cesspit—cover stone, &c.	37
	Glazed stoneware pipe drains	38
	Dip traps or syphon traps	39
	Privy drains	40
	Outlets from soil pipes, &c.	41
	Drain pipes from ditto	42
	Rubble drains	43
	Main drain of brick or sewer	44
	Soil pit	45
Dry area of brick or rubble to prevent damp		46
Water-tank		47
Ventilation under floors		48
Jobbing and fittings		49

RUBBLE MASONRY AND BRICK.

General clause.	Limestone—slate or other material—mortar—bond, &c.	50
Facing.	Random coursed	51
	Regular coursed	52
	Trimmed and coursed	53

INDEX TO MODEL SPECIFICATION. 205

		No
Arched soffits. }	Rough—common brick—hammer-dressed—gauged skew-back	54
Reveals.	Trimmed rough—common brick—neat dressed	55
Arches.	Rough—counter arches—relieving ditto . .	56
	Inverted ditto (see No. 16)	57
Rough projections—cores—corbellings		58
Fire-openings—flues—stacks		59
Coping.	Rough saddle-back	60
	Cement ditto	60
Footing of paving-stone		61
Rough stone worked in walls, &c.		62
Wrought stone built in		63

STONE-CUTTING—CLASSIC, ITALIAN, OR GOTHIC.

Door-steps.	Common	64
	Stair-flights, common	65
	Ditto better	66
	Paving-stone stairs	67
	Out-door steps, plain rubbed	68
Stairs.	Stone stairs, superior moulded	69
	Moulded stairs	70
	Stone steps and iron risers	71
Sills.	Common window-sill	72
	Superior ditto	73
	Gothic sills	74
Dressings, architraves, &c., }	To doors	75
	Ditto and entablatures	76
	Ditto ditto and pediment . . .	77
	Door dressings	78
Window dressings. }	Dressings—architraves—sills, &c.	79
Archivolts.	To doorways or windows—imposts—caps, &c. .	80
Gothic door dressings. }	Plinths—jambs—archivolts	81
Gothic Tudor.	Ditto	82
Gothic window dressings. }	Plinths—jambs—archivolts	83
	Terra-cotta	83
	Superior ditto	84
Tudor ditto.	Ditto	85
Bay or oriel windows. }	Dressings	86
	Balconies	86

	No.
Plinths	87
String course	88
Cornice	89
Blocking course. Parapet—balustrade	90
Chimney stacks	91
Quoin stones	92
Rusticated doors and windows. } Arcades	93
Ashlaring, common	94
Ditto best	95
Ditto fixing	96

Gothic Work.

Window-sills	74
Door dressings	81, 82, 83, 84, 85, 86
Plinth	97
String course	98
Cornice	99
Parapet	100
Chimneys	101
Quoins	102
Ashlaring	103
Buttresses	104
Cappings to ditto	105, 106
Buttress pinnacles	107
Gables	109
Corbels to gables	110

Pediment.	Greek or Italian	108
	Gothic gables	109
	Ditto corbels	110
Portico.	Plinth base to columns—cased	111
	Ditto solid	112
	Back plinth	113
	Columns, &c.	114
	Architrave	115
	Return ditto	116
	Ceiling beams	117
	Stone soffit	118
	Frieze	119
	Cornice	120

INDEX TO MODEL SPECIFICATION.

		No.
Portico.	Blocking—parapet—balustrade	121
	Pediment	122
	Various pavements, steps, &c.	123
Arcades		124
	Gothic	125
Plugs, cramps, lead, &c., to stone-work generally		126

MISCELLANEOUS STONE-WORK.

Coping		127
Curbs		128
Hearths—back and front		129
Chimney-pieces		130
Paving, common		131
	Better	132
	Superior	133
	Marble	134
	Encaustic or tesselated	135
Fittings in cellar		136
Sinks		137
Bath		138
Various fittings		139
Stables, miscellaneous stone-work in		140
Coach-houses ditto		141
Stable-yards, &c., pitch paving, curbs, hinge-stones, &c.		142

SLATING.

Slating, common		143
	Better	144, 145
	Superior	146
	To iron roofs	147
Modes of covering iron roofs		(*note*) 147
Slating to circular roof		148
Flat pitched slating		149
Outside pointing		150
Slate ridges and hips		151, 152
Filleting		153
Queen slating		154
Final clause		155

TILING.

Tiling, plain	156
Ridge and hip tiles	157
Pantiling	158
Final clause	159

Plaster and Cement Work.

	No.
Patent cement, inside work	160
Parian or Keene's cement, ditto	161
Common internal plastering	162
Common three-coat work for ceilings and papering	163
Three-coat work for painting or colour	164
Best three-coat work for paint or paper	165
Whiten ceilings	166
Colour walls, &c.	167
Beads, quirks, &c.	168
Sides, backs, soffits, &c., not cased with joinery	168
Cornices and enrichments	169
Cement skirting	170
Scagliola work and Keene's patent, &c.	171
Patent cement outside brickwork	172
Ditto ditto common, on rubble	173
Ditto ditto superior, on rubble	174
Rough-cast on rubble	175
Two coats common and one Aberthaw, on rubble	176
One coat ditto and two ditto ditto	177
Portland or Aberthaw, on brick	178
Cement or moulded work	179
Parapets and damp-proof course	180

Carpenters' Work.

Inclosures	181
Shoring and old material	182
Piling and planking	183
Sundries	184
Bond and lintels	185
Story-posts	186
Bressummers	187
Quarter partitions	188
Ground joists	189
Common joisting	190
Binders and girders	191
Single-framed floors	192
Double-framed floors	193
Floor trusses	194
Cross straining	195
Floor pugging to prevent sound	195
Ceiling battens	196
Flats	197

INDEX TO MODEL SPECIFICATION.

	No.
Lanterns	198
Roofs, Italian	199
Ditto, Gothic	200
Dormer doors and windows	201
Boarding and battening for lead and slates, and felting	202
Open roof, Gothic	203
Curb roof	204
Garrets	205
Ceiling floors	206
Projecting eaves	207
Troughs, cisterns, &c.	208
Joists, wrought fair	209
Sundry rough work	210
Battening on walls	211
Cradling and firring	212
Columns, and coved ceilings	213
Final clause	214
Sliding doors	215

JOINERS' WORK.

	No.
Ventilator to roof	216
Sky-lights	217
Ceiling or dome inner lights	218
Light and ventilation of water-closets	219
Dormer door	220
Ditto windows	221
Trap door	222
Gutter cornice and cantilevers	223
Ditto Italian	224
Eaves gutter cornice, Gothic	225
Barge boards	226
Ditto Gothic	227
Lanterns	228
Floor boarding, common	229
Ditto better	230
Ditto superior	231, 232
Ditto best	233
Ditto wrought underside	233
Very superior floor of deal and wainscot, or of wainscot wholly	234
Inlaid floors, parquetry ditto	235
Skirting, flush	236
Ditto plugged to wall, common	237, 238
Ditto on fillet and grounds	239, 240
Ditto ditto ditto and grooved into floor boarding	241, 242

	No.
Door, common, ledged and braced	243
Ditto framed, ledged and braced	244
Ditto for coach-houses	245
Ditto panelled, common	246
Ditto ditto better	247
Ditto ditto superior	248
Ditto ditto best	249
Doors, folding	250
Ditto, sliding	251
Outer doors	252
Ditto folding	253
Back doors	254
Ditto folding	255
Doors with side and top lights	256
Ditto side lights only	257
Ditto segment, semicircular, or pointed-headed	258—260
Doors, Gothic	261
Sundry doors, borrowed lights, &c.	262
Lantern light (see 228)	263
Ditto with curved heads to lights	264
Window, sash, simplest	265
Ditto common	266
Ditto better	267
Ditto improved	268
Ditto best	269
Space for Venetian blinds	270
Hinging and furniture of shutters	271
Sash window with lifting shutters	272
Improved pulley stile	273
Ditto superior	274
Ditto triple light	275
Bow sash window	276
Bay sash windows	277
Venetian window	278
Wooden casing pilasters to windows	279
Segment, circular, or pointed heads to windows	280
Casement windows, leaded glazing	281
Ditto *not* leaded	282
If shutters, &c., inside, or outside, &c.	283
If Gothic casements	284
French casements	285
Swing casements	286
Coin beads and window boards	287
Windows with luffer boarding	288

INDEX TO MODEL SPECIFICATION. 211

	No.
Clock or bell-turret	289
Stairs, common	290
Ditto better	291
Ditto best	292
Iron stiffening balusters	293
Fascia of landing	294
Panelled soffit to stairs	295
Inclosure under stairs	296
Cast-iron balusters	297
Step ladder	298
Panelled framed spandrils	299
Casings of carpentry, &c.	300
Board linings, &c., panelled ditto, &c.	301
Wooden columns and pilasters	302
Ditto entablature, &c.	303
Wood carved-work	304
Papier-mâché work, &c.	305
Water-closet fittings	306
Privies	307
Sundry fittings	308
Stables. Miscellaneous joinery	309
Coach-houses. Ditto	310
Loose boxes	311
Outhouses	312
Stable, coach-house, and outhouses, doors and windows	313
Final clause to joinery	314

IRON AND METAL WORK.

Window guard-bars	315
Windows of dairies and larders, fly wire	316
Metal sky-lights	317
Iron chimney bars	318
Binding bolt to hearth arch	319
Iron columns	320
Iron girders	321
Iron joists	322
Fire-proof floors	323
Sundries	324
Iron roofs	325
Cantilever gutters	326
Iron gutters	327
Water-pipes	328
Gratings	329
Fixed ditto	330

	No.
Sundries	331
Rails, balusters, and palisading	332
Iron gates	333
Iron doors	334
Wood and iron doors	335
Iron casements	336
Sundries to windows	337
Iron shutters	338
Iron mangers and racks	339
Cottam and Co.'s stable-fittings	340
Verandah, &c.	341
Sundries	342
General clause for iron work	343
Grates, stoves, ranges, coppers, &c.	344
Bell hanging	345
Heating apparatus	346

PLUMBERS' WORK.

	No.
Lantern top	347
Ridges and hips	348
Dormers, tops	349
Sides of ditto	350
Valleys	351
Chimney gutters	352
Parapet gutters	353
Flats	354
Roofs	355
Flashings	356
Gutter cornices	357
Cisterns and troughs	358
Laying on water	359
Supply-pipes from cistern	360
Water-closets	361
Linings of washing troughs, baths, &c.	362
Pump	363
Sundries	364
General clause	365
Glazier	366
Painter and stainer	367, 368
Zinc worker	369

MODEL SPECIFICATION.

*Specification of works to be done in the construction of . . . . .
. . . . agreeable to the Drawings herewith
furnished, and numbered 1 to . . . inclusive.*

No. 1.
Notice to local authorities.

Payment of fees.

Protective inclosures.
Restoration of pavements.

Clerk of the Works' office.

To give to the Metropolitan Board of Works, local Commissioners and Surveyors, &c., all requisite notices; to obtain all official licenses for temporary obstructions, inclosures, openings into common sewers, water-pipes, &c.; and to pay all proper and legal fees and charges to public officers and neighbouring proprietors, making good any damage occasioned to adjoining premises, and keeping up lights, &c., required by night. Construct proper inclosures and fences for the protection and convenience of the public during the progress of the works; and perfectly reinstate pavements, &c., to the perfect satisfaction of the Town Surveyors.
An office for Clerk of the Works, with fireplace and flue of brick or masonry (see 200).

No. 2.
Removal and shoring.
Old materials.

Carefully take down the old buildings, effectually shoring up as may be necessary the adjoining properties; and entirely remove the old materials, rubbish, &c., to the satisfaction of all parties. The old materials to become the property of the Contractor, who shall be allowed to re-employ only such portions thereof as the Architect under his handwriting shall permit.

No. 3.
Clearing, excavating, refilling.

Clear away all rock, soil, or rubbish, necessary to leave the site of the intended building clear and unincumbered; and excavate for basement story, areas, footings of walls, cesspits, drains, tanks, vaults, &c., as shown by drawings. Properly refill, ram down, and level as required; and remove all superfluous matter excavated, to the satisfaction of all parties.

No. 4.
Drainage of the site before building.

Bale out, draw off, pump away, and remove all water and soil which may come into the excavations from springs, currents, drains, cesspools, rain, or otherwise; and effectually complete the drainage of the excavations and footings before any masonry or brick-work be carried up. Shore up ground as required.

No. 5.
Underpinning.

Underpin in the most careful manner all walls, partitions, or buildings, surrounding the site of the intended new buildings in any way endangered by the excavations of the latter.

No. 6.
Trenches for footings.

Make perfectly level, and hard, the bed of all trenches for footings; and consolidate the earth about the same, and against all walls, drains, pits, &c. The depth of the footings to be contracted for, as shown by the drawings. Should a less depth be admissible, or a greater depth be required, the deviation will be made, under the written permission of the Architect, and accounted for accordingly.

No. 7.
Artificial levels for pavements, and ground-making round building.

To provide, bring in, spread over, and well ram and consolidate, any dry hard ground or rubbish which may be necessary to form the proper level for internal pavements, or paving of courts, areas, pits, &c.; or which may be required to raise the ground level to the lines shown on elevations or sections: the said made ground to extend . . . . . feet from the fronts, and thence to fall to the natural, or now-existing, surface, in an angle of . . degrees.

PILING AND PLANKING. 215

No. 8.
Well digging and building.

To dig a well, in the situation marked on plan, four feet diameter, and . . . . . feet deep below the level of . . . . The same to be properly steined round with . . . . and domed over with . . . .

No. 9.
Concrete foundation.

To make an artificial foundation for (either certain parts of, or) the entire building, ... feet in width, and . . . . . . feet in depth ; the same to be composed of one part of the best fresh quick stone lime, beaten to fine powder, and six parts of unscreened gravel (or fine and coarse stone ballast), mixed thoroughly with each other in small quantities at a time, the lime being moderately slaked with water at the moment of admixture ; and the concrete, when properly compounded, and yet hot, to be thrown from an elevation of not less than ten feet into the trenches, where it will form in layers of six inches deep [or to be rammed in layers] to be repeated one above the other, until the full depth of the required substratum is attained.

[Here specify concrete walling, if any.]

No. 10.
Piling and planking.

To make an artificial foundation for (either certain parts of, or) the footings of entire building, with sound fir piles, ... feet long, ... inches square, pointed with iron, and hooped with ditto ; each pile to be firmly driven by means of a proper apparatus ; the relative situation and distance of the piles, as shown by adjoining figured sketch. Sleepers " × " over each transverse row of piles, and . . . . . planking . . . inches thick ; the whole properly spiked, &c., &c. [For further information as to foundations, see "Foundations and Concrete Works," by Dobson (Lockwood and Co.).]

BRICKWORK.

No. 11.
Indents.

To cut and parget in the old brickwork perpendicular indents to receive the new work, and make good the disturbance in the old work occasioned thereby.

No. 12.
Brickwork.
General clause.

Mortar, &c.

Grouting.
Footings, &c.
See also
No. 69.

The whole of the work, shown by . . . . . tint on plans and sections, to be constructed with . . . . . . bricks, laid in English bond; the said bricks to be the best of their kind, hard-burnt, square, and perfectly sound; laid in mortar, compounded of one-third well-burnt stone lime and two-thirds of clean sharp sand, free from salt, well beaten and worked up together. (See Part VI., Addenda.) No four courses to rise more than one inch beyond the collected height of the bricks. Every course to be filled in and fully flushed up with mortar, and every second (or third course) to be grouted with liquid mortar of hot lime and sand. The footings and walls to be of the varying thicknesses and heights figured on the drawings; and no variation to be made between the outside work and inside work, except that the work intended to be plastered is to have the joints thereof left rough. (See 50.) [Or,

Hollow walls.

The walls to be built hollow with a space of $2\frac{1}{2}''$ or $3''$, the thicknesses being tied together by galvanized iron or tarred cramps set in cement if necessary, and placed every sixth course in height, and about $2'$ $6''$ apart.]

Pointing.

The visible exterior of walls above ground to be finished with a neat flat ruled joint (see 51).

No. 13
Slate or Damp course.

[Lay throughout the length and thickness of all walls and jambs a course of slate in cement, or a layer of pitch or asphalte $\frac{1}{4}$ inch in thickness,

	to prevent the damp rising, 4 lb. lead to be laid over all walls, &c. [The Broomhall Tile Company's patent vitrified perforated damp-proof course is also recommended.]
No. 14 Gauged and other arches, &c.	All front windows and . . . . to have the best gauged arches, abutting on proper skew-backs, the soffits and reveals being 9 or 4½ inches deep. All other outer openings to have plain axed (and slightly cambered) arches closely set and pointed. [*Note.*—A good wall may be formed with facings of brick filled in between with concrete, through courses being placed every four or six courses to bond the work.]
No. 15. Brick rough arches. Relieving arches, &c.	Turn rough arches and counter-arches, wherever practicable, through the entire thickness of walls, except where it may be inexpedient to show them externally (in which case they will be concealed by four-inch facing), and construct nine or four-inch relieving arches, over all lintels or bressummers, as sketch.
No. 16. Inverted arches, &c.	Inverted arches, the whole thickness of walls, under (—external openings,—chimney openings,—and other openings,—from pier to pier,—or) such openings (beneath the ground level) as are shown to have them on the drawings.
No. 17. Brick facing.	To face the visible exterior of the walls of the . . . . with . . . . . . facing bricks of uniform colour, properly bonded into backwork, and finished with a neat flat ruled joint.
No. 18. Brick strings and cutting.	Properly form the string courses, fascias, pilasters, cornices, breaks, recesses, &c., shown by drawings (cutting and rubbing such of the work as may be moulded,) (and neatly splaying angles, plinths, &c.).

[Moulded bricks and terra-cotta are generally used in lieu of cutting, &c.]

No. 19. Brick fireplaces, flues, stacks, &c. Properly form all fire openings, with camber arches over the same, and trimmer arches where required for front hearths. Carefully gather in the chimney throats, and carry up flues of not less than fourteen inches by nine in the clear; well pargetted [or pipe flues of fireclay about eighteen inches long]. The stacks to be carried above roof to the heights shown in drawings, with salient courses, &c.; and properly fix the chimney tops hereafter described.

No. 20. Chimney bars. Put chimney bars of wrought iron, $2\frac{1}{2}'' \times \frac{1}{2}''$, and $18''$ longer than chimney opening, properly caulked at the ends.

No. 21. Tile coping on brick, or brick on edge. The walls of . . . . . . to be finished with a top course of brick on edge [or half-round] (and coped with double plain tile cresting), set in and jointed with new Portland cement and clean sand mixed in equal proportions.

No. 22. Bedding, pointing, and backing. To bed in mortar all the bond timber, plates, lintels, wood bricks, templates, stone and other work requiring to be set in the $\left\{ \begin{array}{c} \text{masonry} \\ \text{brickwork} \end{array} \right\}$. To bed in the point round with lime and hair, mortar all door and window frames, and back up and fill in with solid $\left\{ \begin{array}{c} \text{masonry} \\ \text{brickwork} \end{array} \right\}$ all stone **Fixing grates, &c.** and iron work demanding it. Coppers, stoves, and grates, to be properly set with fire-bricks.
[Form all warm-air flues and fresh-air inlets as shown.]

No. 23. Dwarf masonry or brickwork. Build all dwarf walls, piers, &c., necessary to receive sills of partitions and sleepers or joists of ground floor, as shown on plans [and lay a course of slate in cement, or an asphalte damp-proof course at proper level].

PAVING, VAULTS.   219

No. 24.
Bricknogging.
Bricknog all partitions which are marked on plans with a red hatching, thus: ▓▓▓▓▓

No. 25.
Outer yard Brick paving.
Pave the . . . . . . with hard . . . . . . bricks laid (flat, or on edge, as the case may allow) in mortar; and grout between the joints with liquid mortar. The bricks to be laid according to pattern here sketched, on a good and firm bottom previously prepared.

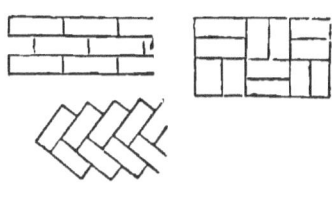

[The "Limmer" asphalte, "Claridge's" and other patent asphalte pavings are preferable.]

No. 26.
Tilepaving, &c.
Pave the . . . . with 12" (red or white paving tiles, laid (square or anglewise) (either of *one* colour or *both*, alternating) in mortar upon full 3" deep of (fine coal ashes, dry brick, stone rubbish, lime core, or concrete) bedding; and the joints thereof pointed with cement.

[Or,

Encaustic Tiles.
Lay on a bed of concrete 3" thick properly floated in cement Minton's or Maw's Encaustic Tiles (here state description or cost per foot).]

No. 27.
Clinker paving.
Pave the (stables, &c.) . . . . . with real Dutch or approved clinkers of approved sample, laid herring-bone fashion or square upon coarse gravel 6" deep, and grouted three times over completely with stone lime and sand. The paving to be laid with proper currents, &c.

No. 28.
Vaults of brick-work or r° masonry.
Construct over the . . . . . (pointed, segment, or semicircular, or elliptical) vaults of $\begin{Bmatrix} \text{r}^\text{o} \text{ masonry } 12 \\ \text{brick-work } 9 \end{Bmatrix}$ inches thick. The spandrils being filled to within 9" of the crown of vaulting with $\begin{Bmatrix} \text{rubble} \\ \text{brick-bats} \end{Bmatrix}$ grouted in liquid mortar.

No. 29. Groined vaults of brick-work or rº masonry.

Construct over the ...... arched and groined vaultings, as drawings (with the groin points [if brick] accurately cut to a regular arris), and the spandrils filled with $\left\{\begin{array}{l}\text{rº masonry brick-}\\ \text{work concrete}\end{array}\right\}$ up to the internal crown of the vault. The whole to be completely grouted with hot liquid mortar; and, after the removal of centering, the whole to be neatly pointed.

No. 30. Corbelled skew-backs.

The skew-backs of vaults to be formed by a corbelling (as c in adjoining sketch), so that the arch does not encroach upon the main substance of the piers or springing walls.

No. 31. Concrete spandrils.

Construct vaults, &c., &c., &c., and fill up spandrils with concrete.

No. 32. Vaults cemented outside.

Construct vaults, &c., and coat the outside of vault and walls with Portland cement ¾″ thick.

No. 33. Lime-whiting.

Stop and lime-whiten twice the ....

No. 34. Iron hooping.

To employ ... cwt. of iron hooping as may be directed, as a bond for the brickwork.

No. 35. Extra brickwork or rubble.

Allow for . . $\left\{\begin{array}{l}\text{perch of extra rubble-work}\\ \text{rod of extra brickwork}\end{array}\right\}$ to be used, or not, as shall appear necessary, and accounted for accordingly.

No. 36. Party-walls.

Party-walls. (Make arrangements with Contractor and adjoining Proprietors.)

No. 37. Cesspit of brick, or masonry.

To construct, where shown on plan, a cesspool, ′ ″ internal diameter, and ′ ″ from the bottom to the springing of (arched or domed) top. The same to be steined round, and vaulted with (4″ brickwork, or good compact rubble masonry), closely pitch-paved, and lined with Portland cement up to the springing of vault.

A man-hole 20" diameter to be left in the top; the same to be covered in with a (Yorkshire, Purbeck, or granite) stone having a strong iron ring therein. Attend Plumber in the admission of water-closet pipes.

No. 38.
Glazed Stone-ware pipes.

[Provide and lay . . . . feet of glazed stone-ware socketted pipes free from fire cracks, &c. (here describe whether of Doulton's or other manufacture, and if with socket caps for cleaning, &c.), laid to a proper fall, and jointed with clay, tarred gaskin, or cement. Provide and fix all proper junctions, syphons, bends, invert blocks, &c., to make connections with sewers.]

No. 39.
Dip-traps or Syphon-traps.

Construct dip-traps where shown on plan; the same to be rendered water-tight with cement; and provide and fix sink stones over the same.

No. 40.
Privy drain.

Put from privy to drain a large and complete brick funnel; or form small trap cesspit under the same, cemented to hold water, and glazed stoneware drain " diameter from thence into larger drain.

[Drains within a building should be avoided if possible, or if necessary they should be enclosed within a brick channel or filled in with asphalte or Portland cement concrete.]

No. 41.
Outlets.

Form at the feet of soil-pipes, waste-pipes, and rain-water pipes, stoneware bends set in Portland cement, delivering into drains [or a trapped open-grated cesspit].

No. 42.
Drain-pipes.

Lay, from the . . . . . to the . . . a drain of strong, glazed, stoneware pipes " diameter clear, and " diameter from the . . . . . . into the . . . . same to be jointed in fine clay or cement.

No. 43.
Drain of rubble, &c.

Construct, and continue, from the . . . . . to the . . . . a drain " by " formed of good close rubble masonry at sides, slate bottom, and strong cover stone. The whole well bedded in mortar.

(Qy. need a *portion* of it be lined with cement?) [These square-built drains superseded. Glazed stoneware pipes can now be obtained of sufficient bore for all purposes.]

No. 44. Main drain of brick or rubble.

Construct, and continue, from the . . . . . . to the . . . . . . a main drain of (brick, or good rubble masonry in mortar) the form and size shown and figured in the annexed sketch. [The invert or bottom of sewer may be formed in concrete, or of invert blocks, and junction blocks of stoneware inserted for inlets. Doulton's hollow stoneware segment sewers are also recommended.]

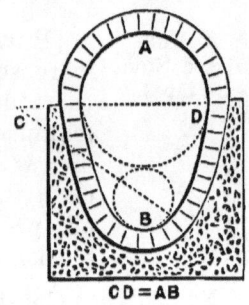

No. 45. Soil or dung pit.

Construct a soil or dung pit, &c., &c. Describe its size, form, material, and what drains it is to receive: whether open; with parapet? arched? with man-hole, &c.? or if covered with slate or stone slabs? whether lined (in part, or wholly) with cement, to hold liquid manure.

No. 46. Dry area of brick or rubble, &c.

Form a dry area round the walls of . . . . . (as shown on drawings), of (brick, or stone rubble-work) and of the sectional form and size shown and figured in the annexed sketch; the same to be (covered with flat stone, or arched with brick or rubble). Man-holes where shown on drawings; and provide and fix stone curb and gratings of iron therein, 20" square. The bottom to have a fall to the drain, and to be pitch-paved.

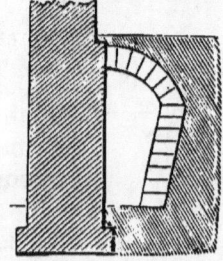

RUBBLE MASONRY. 223

No. 47.
Water-tank,
brick or rubble.

Construct a water-tank below the floor of ..... as shown on plan, and of the sectional form and size here shown and figured. The same to be covered with a semicircular 4" brick arch, having a man-hole, with stone therein, 20" diameter, and iron ring. The sides to be of " (rubble, or brick-work), the bottom, of (ditto), and forming an inverted segment.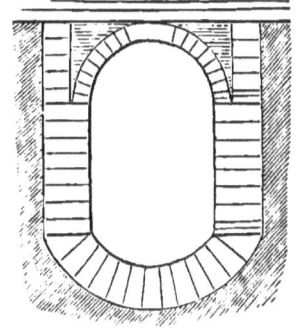
The ground outside the sides and below the bottom to be thoroughly rammed and consolidated; the outside of said sides and bottom to be laid against a clayed backing, and the inside to be lined with fresh Portland cement.

No. 48.
Ventilation.

Leave proper and sufficient openings for ventilating under the joists of ground floor, and provide and fix neat gratings [of cast iron or glazed earthenware] in said openings. Form air flues, where shown, in thickness of walls.
(See Sanitary Construction, Part V.)

No. 49.
Jobbing, and

Attend upon the Stone-masons, Carpenters, Plumbers, and Smith, aiding, and making good after them, and to perform all jobbing necessary to the perfect completion of the works.

Brick fittings.

Half-brick piers to stone or slate shelves of cellar, dairy, larder, &c. (See 148.)

RUBBLE MASONRY AND BRICK.

No. 50.
Rubble masonry.

The whole of the work, shown by . . . tint, on plans and sections, to be constructed of good
{ lime-stone,
slate,
or other } rubble masonry, properly bedded

Mortar.

in mortar, compounded of one-third well-burnt

stone lime and two-thirds of clean sharp sand, free from salt, well beaten and worked up together.

Footings. (See also No. 69.) The footings of walls (see No. 69) to be formed of large flat stones, (having their length not less than the width of the masonry above,) laid transversely, as shown by sketch, A being the footing stone.

Bond and quoins. A sufficiency of bond stones, as B B B in the annexed figure (having an excess of *length* only, and *not* of *height*), at all quoins, and where else required to bind the work, and insure its uniform compactness, especial care being taken to make the walls of equal solidity all through by well filling the inner part with small stones and mortar; and the work to be grouted

Grouting. with hot lime and sand at every rise of . . . inches. The stones to be bedded as found in the quarry. The walling to be carried up, and preserved, both vertically and horizontally true, and of the varying heights and thicknesses shown or figured on the drawings.

No. 51. Random coursed facing. The visible exterior of walls above ground to be finished in neat random coursed work, the stones being hammer-dressed to a fair surface and neat joint, and well pointed.

No. 52. Regular coursed facing, or coursed rubble. The visible exterior of walls above ground to be finished in neat and regu-coursed work; no course to be more than . . . inches, nor less than . . . inches high; hammer dressed to a fair surface; the joints to be close and true, both vertically and horizontally, and pointed with Aberthaw mortar.

RUBBLE WORK. 225

No. 53.
Trimmed and coursed facing.

The visible exterior of the ..... walls above ground to be faced with a neatly trimmed ashlaring of ...... stone, in courses of equal height, nor less than inches, not more than inches. The vertical joints tooled close, and the horizontal joints bevelled (" by ") so as to throw the water from the top of each course, and pointed with (qy. Aberthaw?) a neat flat ruled joint. (Qy. should this ashlaring be *bedded* and *backed* with Aberthaw?)
[The joints may be all square and close.]

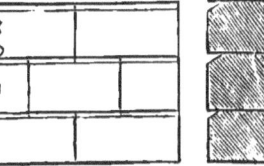

No. 54.
Arched soffits.

All the ...... openings to have—(trimmed $\begin{Bmatrix} \text{segment} \\ \text{semicircular} \end{Bmatrix}$ rough arches)—(good common brick arches)—(neatly hammer-dressed arches)—(gauged arches corresponding with the ashlaring) to form their soffits outside wood frames; the soffits of windows being $\begin{Bmatrix} 9 \text{ or} \\ 4\frac{1}{2} \end{Bmatrix}$ inches deep; and proper skew-backs being formed in all cases. All such arches to be close set and pointed. [Stone or terra-cotta skew-backs and lintels may be substituted.]

No. 55.
Reveals.

All reveals of the said openings to be—(trimmed to a neat face)—(also of brick)—(neatly dressed as arches)—(wrought to a neat sharp arris, as soffits).

No. 56.
Rough arches.

Turn rough arches and counter-arches wherever practicable, through the thickness of walls, excepting where they may not show in the external facing; and construct 12-inch rough relieving arches over all lintels, or bressummers, as sketch.

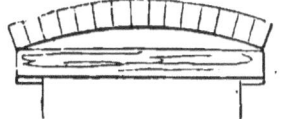

No. 57. Inverted arches, &c. (See No. 15.)

L 3

No. 58. Rough projections.	Properly form all rough projections, cores, corbellings, &c., for cement string-courses, fascias, pilasters, cornices, &c. Or, see No. 18, altering the word "brick" for . . . . . .
No. 59. Fire openings and flues. Chimney stacks.	Properly form all fire openings, with brick arches over the same; and brick trimmer arches where required for front hearths. Carefully gather in the chimney throats, and carry up flues round cylinders of not less than 9" or 12" diameter in the clear, well pargetted. The chimney stacks above roof to be carried up in brick to the heights shown in drawing; and properly fix the chimney heads or pots hereafter described. (See No. 20.)
No. 60. Rough coping.	The walls of . . . . . to be finished with a top course of large rough stones, partially hammer-dressed to a circular top edge or saddle back; bedded on their *flat* edge, and well flush pointed with Aberthaw mortar [or cement].
Rough coping and cement.	The walls of . . . . . to be finished with a top course of stone on edge, well bedded and jointed; to overhang the faces of wall 2 inches, and (when the masonry shall have perfectly settled) to cover the said coping with Portland cement, as adjoining sketch.

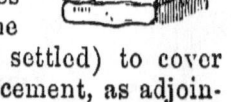

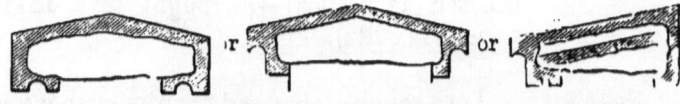

No. 61. Footing of paving stone.	Provide and lay beneath the footings of . . . two complete courses of (Yorkshire) stone, of the several widths shown on drawings. The stones to be 3" thick, each averaging 10 ft., and none less than 6 ft. superficial.
No. 62. Rough stonework in masonry or brickwork.	Provide and fix—(here mention and describe the form and size of any *rough* or *roughly wrought* stonework which has to be worked into the

STONE-WORK.

brickwork or masonry,—such as corbels for overhanging chimney breasts, or other masonry; for girders or other timbers; rough lintels for windows or doors where flat arches are not practicable, and which are to be plastered; rough lintels over intercolumns; rough plinths to receive iron or wood columns and story-posts; rough templates to receive iron beams; &c.) (in short, all stone-work that is to be hereafter concealed).

No. 63. Wrought stonework built into masonry, &c., or brick-work.
Provide and fix—(here mention and describe the form and size of any *wrought-fair* stone-work which has to be worked into the solid brick-work or masonry at the time of its building,— such as hinge-stones, lintels, solid plinths, bases, corbels, &c., which are unconnected with any other stone-work, uncovered by plaster, and used in plain buildings, which, in all other respects, are of common brick-work or rubble masonry). (Example.—The hinge-stones and lintel of a strong closet rebated for iron-doors; the plinths under the piers of a shop front; &c.) (In short, all stone-work which does not partake of the nature of ashlar; which cannot, like stone steps or window-sills, be worked in after the masonry or brick-work has been carried up; and without the previous fixing of which the common walling cannot in any degree proceed.)

STONE-WORK.

No. 64. Door-steps, common.
To put to the . . . doors . . . . . as shown on plan, plain solid tooled steps of $\left\{ \begin{array}{l} \text{granite} \\ \text{Purbeck} \end{array} \right\}$ stone 12″ × 8″, properly back-jointed, and mortised for door-posts. Also a piece of paving of the same, to extend from step to outside face of plinth.

No. 65. Common stair flights.
To put from the . . . . to the . . . . a flight of solid $\left\{ \begin{array}{l} \text{granite} \\ \text{Purbeck} \end{array} \right\}$ steps, 12″ × 8″, properly tooled (and mortised for iron balusters), back

jointed, and securely (bedded on { masonry / brick-work } or (pinned into walls). (If there are landings, describe them.)

No. 66. Better stair flights. To put from the .... to the .... a flight of solid { Purbeck / Portland } steps, wrought and rubbed smooth on all faces and soffit; back-jointed as sketch, the treads mortised for balusters, and the steps securely pinned into walls. Landings of the same inches thick, rebated on to last riser (and, if required, to have joggled joints run with lead).

No. 67. Stairs of paving stone. To put from the .... to the .... a flight of risers, treads, and landings, of inch tooled .... stone, (securely pinned into walls) or (bedded on brickwork or masonry). The treads mortised for balusters.

No. 68. Out-door steps, &c., better quality. To put to the doors of .... solid wrought and rubbed Portland stone steps (qy. with moulded nosings). Each step, out of a stone inches by .. inches, properly jointed and bedded on the substructure (and flanked with Portland curbs, .. inches by .. inches, wrought, with rounded top, rubbed, and properly mortised for balusters). (If landing, state it.)

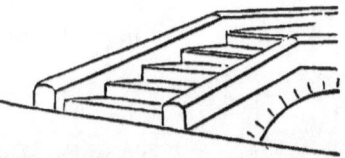

No. 69. Stone stairs, superior. To put from the .... to the .... flights of the best and hardest Portland stone steps, with moulded nosings, returned at ends, the soffits (as well as the rest) wrought and rubbed fair

STEPS AND STAIRS. 229

bead flush-jointed, as sketch, and tailed full
9 inches into the walls; or resting on corbelling.
The bottom step having
its section a solid square,
and finished with hand-
some curtail. The land-
ings thereof to be full
inches thick, with edge
moulding to correspond with nosings of steps;
the mid-landings being in one slab each, and the
upper landing of . . . stones, tailed inches into
walls, and joggle-jointed with lead. Each step
and landing mortised for balusters. [In buildings
exposed to risk of fire, stone is apt to crack and
splinter under great heat; concrete steps are
here preferable, carried on corbelling or iron
carriages.]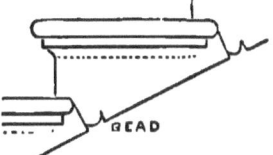

No. 70.
Stone stairs,
handsome.

To put from the . . . . to the . . . . flights of
the best and hardest
Portland stone steps,
with moulded nos-
ings along the front,
outer end, and also
along the back of
each step; the soffit of each step (except those
at bottom and landings) to be moulded as shown
by sectional profile, and the whole tailed 9
inches into walls. Landings . . inches thick,
having their soffits moulded and panelled as
drawings, and their edges moulded to correspond with the nosings of steps. Handsome
curtail to bottom step, whose section will be a
solid square. The whole lapped and jointed as
drawings. The mid-landings to be in one slab
each, and the upper landing in . . . . stones,
tailed inches into walls, and joggle-jointed with
lead. All required holes for balusters.

No. 71.
Stone steps
and iron
risers.

To put from . . . . . to . . . . . a flight of
Portland ⎫
Purbeck   ⎬ stone treads, with (rounded) (or
Yorkshire ⎭

| No. 72. Window-sills, common. | moulded) nosings; the risers thereof to be of open cast iron-work (see Smith), &c., &c. To put to the windows of . . . . . good common sills of . . . . stone, inches by inches; sunk, weathered and throated, and 4 inches longer than the width of openings. 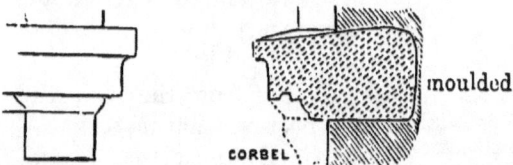 |

| No. 73. Window-sills, superior. | To put to windows of . . . . finely wrought (and rubbed) sills of . . . . stone, inches by inches, (qy. moulded as drawings?) and sunk, weathered and throated. The sills to be 4 inches longer than the (width of openings) (or than the united width of openings and jamb dressings). (Qy. whether corbels under the sills. &c. ?) |

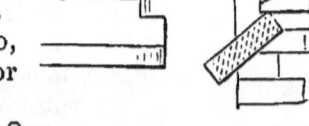

moulded.

| No. 74. Window-sills, common Gothic. | To put to windows of . . . . common sills of (slate, Yorkshire stone, or Purbeck) paving stone, inches wide, 2½ or 3 inches thick, wrought-fair edge and ends, and laid sloping. (The above will do, either for Italian or Gothic.) |

Or

| Window-sills, Gothic. | to put to windows of . . . . . . sills of . . . . . stone, properly wrought, (throated, if projecting;) (specify if in one or more heights, and make the *top* sill of one length, if possible). Note, if two sills may be cut out of one stone, as sketch. |

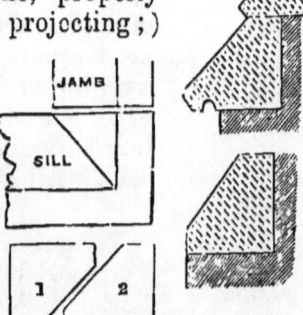

Or

[put to windows of . . . . moulded brick or terra-cotta sills of the section in margin, also lintels as shown.]

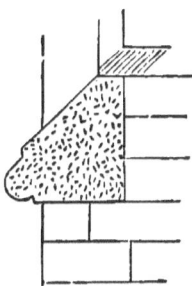

No. 75.
Architraves to doors.

To put to the doorway of . . . . an architrave of the best solid . . . . . stone, moulded, and of the scantling shown by fig. drawing No. . The lintel to be of one stone; the jambs of one three or more } stone(s) down to plinth. [It is a fallacy to imagine one stone stronger than several. Constructively it is better to have several joints in a great length of jamb or mullion, especially where brick walling is used, the settlement of work being 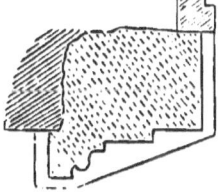 more uniform and the weight of superincumbent masonry does not bear entirely on the jamb or mullion, causing fracture at joints, &c.] The whole to be rebated, as drawings. (State if any of the jamb stones are to bond into the walls.)

No. 76.
Door architrave, and entablature.

To put to the doorway of . . . . an architrave, &c. (See No. 75.) Put over the architrave a frieze of similar stone and quality in . . . . piece(s), not less than inches thick, of the height shown in drawing; and, above the frieze, a cornice of the sectional scantling and moulded profile also shown (and in . . . . stone(s).

No. 77.
Door architrave, entablature, and pediment.

To put to the doorway of . . . . (all that is mentioned in Nos. 75 and 76) a pediment of the same stone and quality, the tympanum of one (or three) stone(s) not less than   inches thick, and the raking cornices to correspond with that below, having the additional moulding shown on drawing.

No. 78.
Door dressings, various and handsome.

To put to the doorway of . . . . pilasters, architraves, &c., &c., &c.,

or

engaged columns, architraves, &c.,

or

pilaster dressings, as drawings,

or

columns and jambs, as drawings,

with

moulded bases, and capitals,

or with

Ionic, Corinthian, or enriched capitals,

or with

consoles plain or enriched, as drawings:

the pilasters, engaged columns, &c., to be of the best solid . . . . stone, in . . . . piece(s) (exclusive of caps and bases) and of the sectional form and scantling shown on drawings. The architraves, &c. (see Nos. 75 and 76), and, if pediments are required (see No. 77).

Or,

If any other parts of the building are

the capitals (and, if there be such, the enrichments, of architraves, friezes, cornices, consoles,

to be enriched, this may be reserved till the end as a general clause.	&c.) to be executed in the very best style, after models of the full size, provided at the cost of the Contractor, and only adopted under the expressed satisfaction of the Architect.

Or

Moulded brick dressings.	[instead of stone, moulded brick pilasters, architraves, columns, &c., may be employed. In this case state manufacturer's name and give detail of the same.
	Stone or concrete heads or lintels to windows and doorways may be substituted for architrave dressings or pediments, the soffits being simply splayed, or moulded, or of segmental or pointed shape.] (See sketch.)
No. 79. Window dressings, various.	To put to the window openings of the . . . . . . architraves, &c. (See No. 75, substituting the word "sill," or "blocking course," or "string course," for plinth) (or as the case may be.)
	[The dressings specified for doors may be here repeated, describing any variations.]
No. 80. Archivolts, &c. Imposts, &c., &c., &c.	To put to doorway of . . . . . or the windows of . . . . . an archivolt, or archivolts of . . . . . the best solid stone, moulded, and of the scantling shown in drawing No. , with joints only, as shown on the elevation thereof by blue lines. (If a key-stone, describe it.) The imposts to said archivolt of similar material and quality, and of the substance and profile shown in drawings. (Qy. forming caps to pilasters or jambs? which pilasters or jambs will extend in . . . . piece(s) to base, plinth, sill, string or blocking course, and be of the sectional substance shown on plan.)

No. 81. Gothic door dressings.	To put to doorway of . . . . . a plinth, jambs, and archivolts of solid . . . . . stone, wrought, moulded, and of the sectional form and scantling shown by drawing No. .

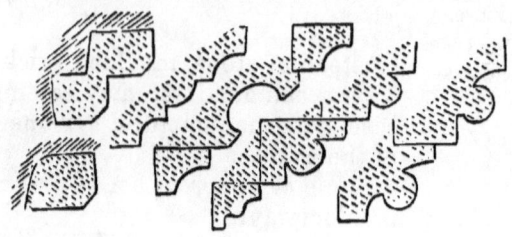

The jambs to be of . . . . . stone(s), alternately bonding into walls; the archivolts of . . . . stone(s); and the same to be of . . . . . stone(s) in their recessed depth, as indicated in sections. (If there be imposts, caps, bases, plain-moulded or carved, other enrichments, and label or drip-stones, state them.)

No. 82. Gothic Tudor door dressings.	To put to doorway of . . . . . a plinth and jambs of solid stone, with square head and spandrils inclosing an archivolt, wrought, moulded, &c., and of the sectional forms and scantling shown by drawing No. . (See No. 81, and add thereto a description of the spandrils.)
No. 83. Gothic window dressings.	To put to window openings of . . . . . jambs and archivolts of solid . . . . . stone. (See No. 81.)
Moulded terra-cotta.	[In any of the above cases moulded bricks or terra-cotta may be used instead of stone.]
No. 84. Gothic windows, superior.	To put to window openings of . . . . jambs and archivolts of solid . . . . stone (see No. 81), and properly cut, carve, and fix the (columns, mullions, transoms, mullion arches (plain or foliated), spandrils, and the tracery complete), as shown on drawings. The (columns, mullions, &c.) to be of the sectional form and scantling shown in details, and to have

GOTHIC DRESSINGS.

joints only where marked by blue line on elevation.

No. 85. Gothic Tudor windows. To put to window openings of . . . . . . jambs and square head of solid . . . . stone, (qy. inclosing mullions, transoms, mullion arches (plain or foliated), spandrils, and tracery) complete, as shown on elevation. The jambs, heads, (mullions, &c.), to have the sectional form and scantling shown in details, and to have joints only where marked. Properly cut and fix also the labels or drip-stones.

No. 86. *Note.* Bay or Oriel; Italian or Gothic. Bay or Oriel windows, Italian or Gothic, will partake of the same *general description* as already given, to which it will be necessary to add a description of the plinth (under the sill), the angular piers or jambs, the blocking course, balustrading, &c. (if Italian): or the cornice and battlemented or pierced parapet (if Gothic). It may be, also, that the Gothic oriel may rest on a moulded corbel (which must be accurately described as to construction); and that the **Balconies.** Italian windows may have balconies before them (continuous, or attached separately), in which case they must be described, as formed of . . . . . . stone landing, inches thick (how wrought and moulded?), tailed inches into wall; supported by carved brackets or consoles (as drawings) securely pinned inches into wall; and supporting a blocking course, with pedestals, balusters, capping, &c., wrought moulded, &c., &c., as drawings.

[Terra-cotta or cast-iron balustrade may be employed in lieu of stone.]

No. 87. Plinth. To put along the . . . . . a plinth of . . . . . stone (neatly wrought) (wrought fine) (how tooled?) (qy. rubbed?) . . feet . . inches high, in

{ one / two / or more } stone(s) inches thick. The top chamfered, and no stone to be less than .. feet .. inches long. (See No. 97.)

No. 88. String course. To put along the ..... a { wrought moulded / plain wrought } chamfered and throated (say what stone?) string course, of the sectional form and scantling shown in details. No stone to be less than   feet in length. (See No. 98.) [Very effective terra-cotta string courses may be employed.]

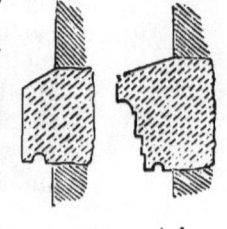

No. 89. Cornice. To put along the ..... a cornice of ..... stone (qy. in { two / or more } layers), of the sectional form and scantling shown in details, and no stone to be less than   feet in length. (Enrichments to certain mouldings? Modillions, plain or enriched? Soffits sunk panelled, &c.? Dentils? Antefixæ?) All the said enriched parts to be carved in the best style out of the solid, after models of the full size, provided at the cost of the Contractor, and only adopted under the expressed satisfaction of the Architect. (See No. 78.) The plain parts wrought in the best manner, (qy.) { and rubbed? / or tooled.

No. 90. Blocking course. To fix above the cornice, all along the ......, a blocking course of ..... stone,

or

Parapet. a parapet having a plinth, dado, and capping,

BALUSTRADE, QUOIN STONES. 237

or

Balustrade. a balustrade course, composed of plinth, pedestals, balusters, and capping, as drawing; the same (or the several parts of the same) to be solid, and of the sectional form and scantling shown in detailed drawings. The plinth and capping to be in stones of not less than   feet long.

The balusters $\begin{cases} \text{square} \\ \text{turned} \\ \text{or otherwise} \end{cases}$ and half-balusters to pedestals. The whole to be (wrought fair) (tooled ?) (rubbed ?) (See No. 100.)

[In the generality of cases it is preferable to let the roofing run over the cornice and to form a sunk gutter, or to provide a moulded cast-iron one. See sketch.]

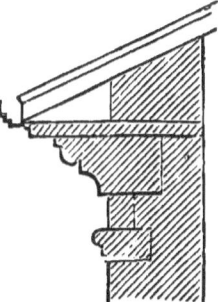

No. 91. Chimney stacks.
The chimney stacks of . . . . to be capped with . . . . . stone, moulded, and of the sectional form and scantling shown in drawing. (Qy. whether the shafts shall be also of stone entire; or stone ashlar.) (See No. 101.)

No. 92. Quoin stones.
The . . . . angles of . . . . to be finished with (roughly wrought) or (wrought, part rough, part tooled) or (wrought fair) and chamfer channelled quoins of solid . . . . . . stone, of the heights, and lengths, sectional form shown and figured on detailed drawing, two quoins being cut out of one stone thus:

(See No. 114.)

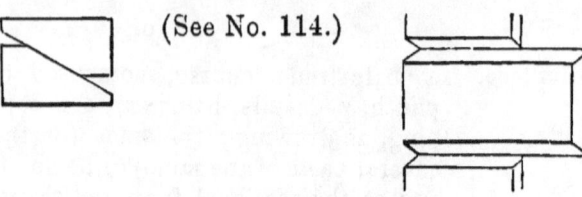

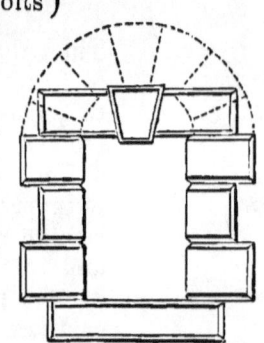

No. 93.
Rusticated doors and windows, and arcades.

The $\left\{\begin{array}{ll}\text{jambs} & \text{heads} \\ \text{reveals and soffits} \\ \text{sills} & \text{archivolts}\end{array}\right\}$ of the (arches) (doors) or (windows) of . . . . to be formed with rustic-work of solid stone, corresponding, in material and workmanship, with the rustic quoins of main building; and of the form, size, and scantling shown on detailed drawings.

No. 94.
Ashlaring, common and superior.

To face with . . . . . stone ashlaring, (roughly wrought) (wrought neat and tooled) (wrought fair and rubbed) and (square- chamfer- or moulded-) channelled, as shown by elevation, the whole of the . . . . . , properly forming the radiating channels of $\left\{\begin{array}{l}\text{flat} \\ \text{segment} \\ \text{circular}\end{array}\right\}$ arches, and returning the channelling under soffits and against jambs or reveals of doors, windows, (recesses?) and all round (qy. insulated piers?). (See No. 96.)

Or

No. 95.
Ashlaring, the best.

to face with . . . . . . stone ashlaring, wrought fair (qy. rubbed) and worked close joint, the whole of the . . . . . . (If there be no *architraves*, mention the radiating joints of $\left.\begin{array}{l}\text{flat} \\ \text{segment} \\ \text{circular}\end{array}\right\}$ arches, &c.) shown by elevation(s). (See Nos. 96 and 103.)

STONE ASHLARING.

No. 96.
Ashlaring, how fixed, &c., &c.

The said ashlaring to consist, as nearly as circumstances will admit, of courses ' " high; formed, with headers, having a horizontal bed of ' " by ' "; and stretchers, having a horizontal bed of ' " by ' "; the quoin stones being in no direction less than " on their bed. All horizontal joints to have a slight chamfer on the upper edge of each stone. (See No. 103.)

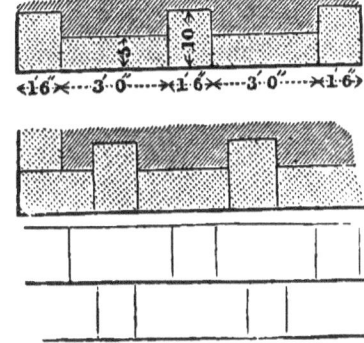

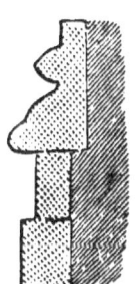

No. 97.
Gothic plinth.

Gothic basement or plinth. For general description, see No. 87. Add description of moulding and subplinth (if required), ", the same to be of the sectional form, &c., shown in details.

No. 98.
String course, Gothic.

Gothic string course. Same description as No. 88.

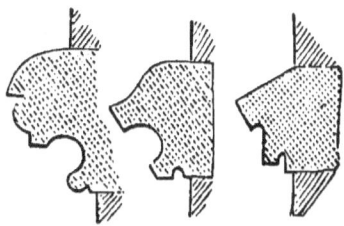

No. 99.
Cornice, Gothic.

Gothic cornice. Same general description as

240   HINTS TO YOUNG ARCHITECTS.

No. 89. Specify whether plain? with separate, or continuous enrichments?

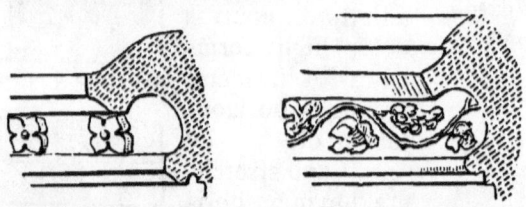

No. 100.
Parapet,
Gothic.

Gothic parapet. To fix above the cornice (or string course) all along the . . . . a

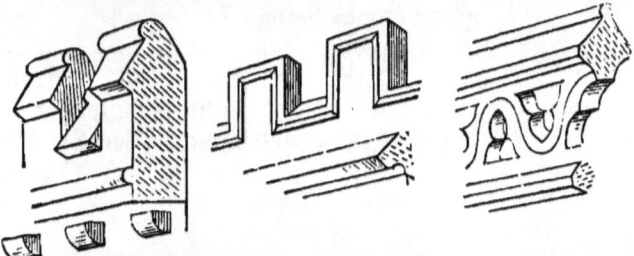

(plain capped) or (embattled and capped) or (embattled and moulded) or (open worked and capped) parapet of . . . . stone, as drawings. The several parts of the same to be of the sectional form and scantling shown by details (the capping, if continuous, in stones of not less than ′ ″ long) (the open or sunk ornamental work carved in the best manner), and the work, generally, to be jointed as shown by blue lines on detailed drawing.

No. 101.
Chimneys,
Gothic.

Chimney stacks, with (shafts) caps, bases, plinths, &c., (plain moulded) (em-

BUTTRESSES. 241

battled) (panelled) (or otherwise decorated) as drawings; the same to be of . . . stone, of the sectional form and scantling shown by details.

(Qy. whether the *plain* parts of the plinths, and shafts, may not be of brick; the *moulded work only* being of stone?)

[Terra-cotta chimney stacks may be used instead of stone.]

No. 102.
Gothic quoins.

Gothic quoins. Same general description as No. 92.

Ditto to doors, windows, &c. (See No. 93.)

No. 103.
Gothic ashlar.

Gothic ashlaring. Same general description as Nos. 95 and 96.

No. 104.
Buttresses.

Describe buttresses. Whether formed with

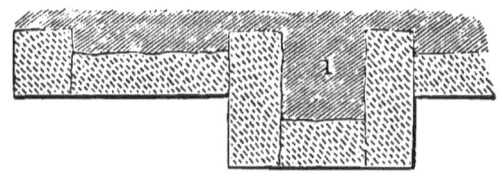

ashlar, (as 1)? or with solid work (as 2)? or part solid and part ashlar? or with heading and

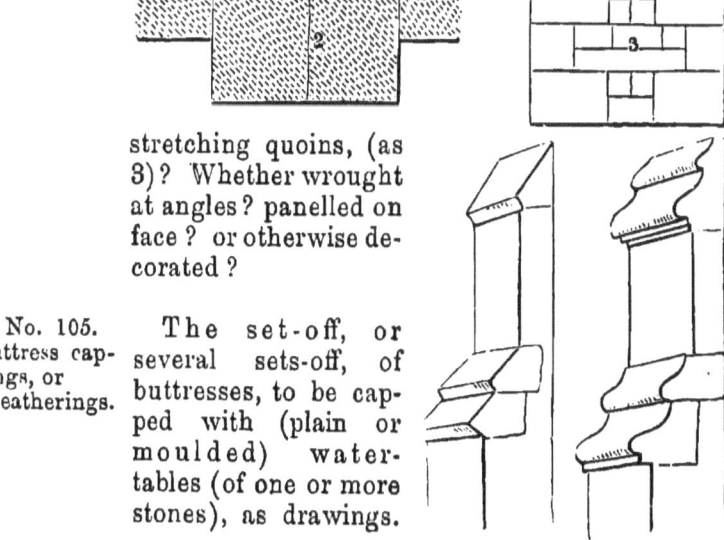

stretching quoins, (as 3)? Whether wrought at angles? panelled on face? or otherwise decorated?

No. 105.
Buttress cappings, or Weatherings.

The set-off, or several sets-off, of buttresses, to be capped with (plain or moulded) watertables (of one or more stones), as drawings.

M

The same to be of the same material and quality as . . . and of the sectional profile and scantling shown in details.

Or

No. 106.
Gablet capping.

(same general description as No. 105), substituting for "water-tables" gablets (plain moulded) (topped with finials) (with carved finials and crockets), &c.

Or

No. 107.
Pinnacles, &c.

(same general description as No. 105), substituting for "water-tables" gablets (as described in No. 106) (and adding,)

The top gablet to be crowned with (plain moulded) (moulded crocketed) (panelled) pinnacles, having carved finials, &c.

[*Note.*—The Gothic details here given are somewhat effete; crockets and other exuberances of the style are best avoided in this climate in exposed situations. See Principles of Design-Stonework.]

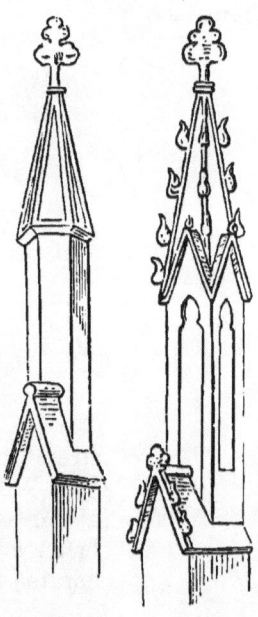

No. 108.
Pediment, Greek or Italian.

The ashlaring in the tympanum of pediment to be precisely accordant, in the height of its courses, and construction, with that of the (walls below, or general face of building). The horizontal cornice to be a continuation of (main

PEDIMENTS. 243

cornice), omitting the top moulding; and the
upper layer thereof to be of single stones from

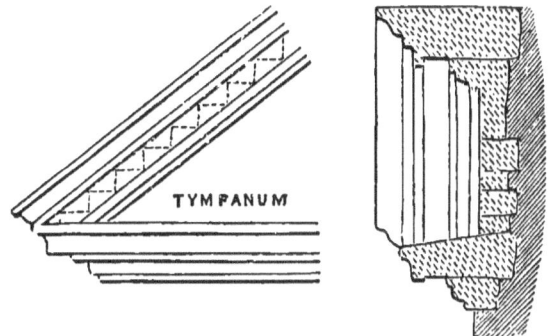

front to back, securely tailed into the masonry
not less than   inches, and no
stone having a front length of
less than ′ ″. The raking
cornices to be of the same form
and scantling as that of (main
cornice). The apex or meeting
mouldings at top to be out of
one block, having a horizontal
bed on tympanum; and the
raking cornice at the lower
angles of pediment to be out of
the same undivided block with
the end of horizontal cornice.
The hidden part of raking
cornices to be cut in the
form of steps, so as to
have a series of horizontal
beds upon the back ma-
sonry. The top stones of
raking cornices

either thus,   or thus,

   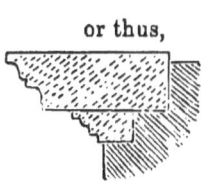

to be of one piece transversely; and no stone
having a front length of less than ′ ″.

M 2

If blocks, pedestals, or acroteria, describe them.

No. 109. Gables, Gothic.

The gables of . . . . . to be capped with a (plain) (moulded) coping of . . . . stone, of the

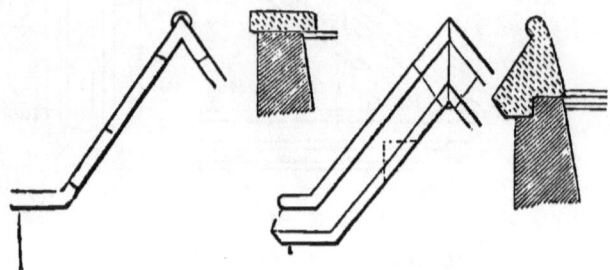

sectional form and scantling shown by detailed drawings, in lengths of not less than ′ ″, (back-notched for horizontal beddings,) and with springing stones and apex-saddle stones cut of the solid, as also shown.

No. 110. Gothic gable corbels.

(The springing stones to be supported by cut flush corbels, of the face and profile shown by drawings;)

or

(the springing stones to have a return face, supported by corbels of the face and profile shown by drawings) (the said corbels to be of *stone, serving to stop the eaves cornice or gutter.*)

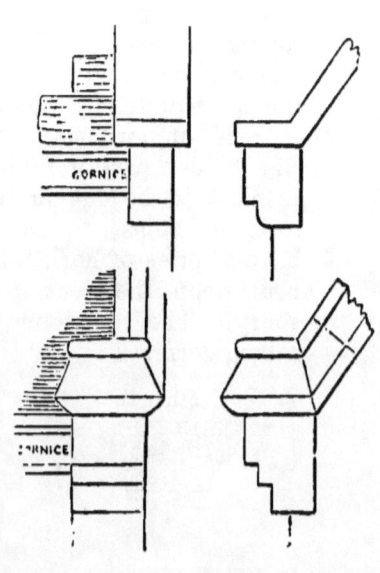

Portico. See Note below.

Portico, Greek or Italian.
No. 111. Plinth, cased.
The plinth under the columns of the portico to be formed of top and side casing of . . . . . stone, of the sectional form and scantling, and vertical jointing, shown in 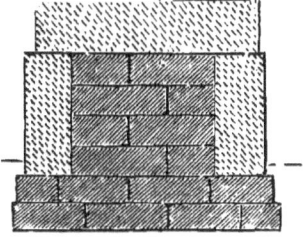 drawings, properly bedded on the (brick, or rubble) basement and core. The sub-plinth bonded into said core, having its bonding stones under the axes of columns ;

or

No. 112. Plinth, solid.
the plinth under the columns of portico to be formed solid, of . . . . . stone, of the sectional form and scantling,  and vertical jointing, shown on drawings ;

and

No. 113. Back plinth.
a plinth, of ashlaring, to match the stone-work under columns, to be carried round the inside recess, or back of portico, as drawing.

No. 114. Columns, &c.
The columns (antæ) and pilasters to be of . . . . stone, with (moulded, or moulded and enriched) bases and capitals, and (plain, or fluted) shafts, as detailed on drawings. The shafts to be in (one stone, or three stones), and the pilasters properly bonded into the main walling.
[A layer of felt, sheet-lead, &c., is sometimes interposed between the beds when there is great weight of superstructure or tendency to unequal settlement.]

Note.

Portico
The portico will either be constructed with a substructure of common rubble, or brick ; a

plinth of brick, or rubble covered with cement; columns of brick covered with cement; an entablature, &c., of rubble or brick (with rough stone lintels over columns), also cemented;

or

the visible portions will be *partly* stone, as
    1st. Stone plinth only;
    2nd. Stone plinth and columns;
    3rd. Stone plinth, columns, and architrave;

or

the visible portions will be *wholly* stone, with backings and fillings of rubble or brick, as the locality may require.

It will therefore be necessary,
    under the heads of 'Bricklayer,'
    or 'Rubble Mason and Bricklayer,'
to describe the foundations and the *core* of the work; whether there are to be flat *stone footings, inverted arches* under the columns, *wood bonds and cores* to the brick columns, *relieving arches* over the same, &c., &c.*

Arcade.   The same remarks will also apply to arcades. Such portions, therefore, as are not to be stone, will be described under the heads of 'Bricklayer,' or 'Rubble Mason and Bricklayer.'

No. 115. Architrave.   The architrave to be of . . . . . stone (solid)¹ (or solid up the first, or first two faces)² (or solid³ up to the crown moulding)⁴ (*describe the casing to the part which is* NOT *solid*), of the sectional form and size shown by details, and *vertically* (or

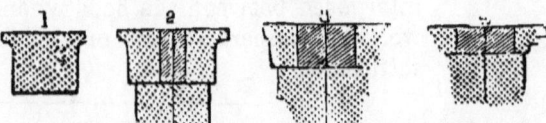

otherwise) jointed, as marked on elevations, or

---

* See Bartholomew's Specifications, No. 4600.

ARCHITRAVES. 247

shown or described on detailed drawings. If enriched, describe it.

Note.—It is impossible to make any general description sufficiently accurate for this important member of a colonnade. A reference to a fully detailed drawing, showing the stones separately, the mode of uniting them by arched or vertical joggled joints, the copper chain tie and hanging bar, and the relieving arches of the concealed brick-work or masonry, is the only way of insuring a clear understanding.*

No. 116.
Return or back architrave.

An architrave of ashlaring, to match that over the columns, to be carried (round and) along the inside (or back) *recess* of portico, as drawing. Qy. enriched?

No. 117.
Beams in ceiling.

If there be any inner longitudinal and transverse beams to form the ceiling of portico, they must be carefully studied, and here described. Qy. enriched?

No. 118.
Stone soffit.

If there be a stone ceiling altogether, here describe it. Qy. enriched?

No. 119.
Frieze.

The frieze to be formed of . . . . . stone ashlaring, in no case less than   inches thick, jointed as drawings; and the quoins to be cut out of solid stone, so as to show a return of not less than ′ ″. Qy. enriched?

No. 120.
Cornice.

To put along the front and returns of portico a cornice of . . . . . stone. (See No. 89.) Qy. enriched?

No. 121.
Blocking, parapet, balustrade.

If the portico, instead of the pediment, &c., is to have a plain blocking course—or parapet with capping and plinth—or open balustrade (see No. 90). Qy. enriched?

---

* See Bartholomew's Specifications, No. 4610.

No. 122. Pediment.	Adopt the general description given at No. 108. If the portico be surmounted by blocks, pedestals, or acroteria, describe them. Qy. enriched?
No. 123. Various.	Complete the description of the portico by explicit references to its landing, pavement, steps, guard-stones to preserve the plinth from carriage-wheels, &c.
No. 124. Arcades. Roman.	Describe the plinths; whether solid or not. The piers; whether solid or of ashlar; whether plain or rusticated (see No. 92); whether there be plain or moulded imposts. The arches; whether with archivolts, or radiating stones, plain, or rusticated as piers; whether key-stones, plain or carved, &c., &c.
No. 125. Arcades, Gothic.	Describe the plinths; whether solid or not. The pillars (their bases, capitals, if any) and the number of stones to compose the shafts; the number of stones in the archivolts; and the quality of the work filling up the spandrils.
No. 126. Plugs, cramps, and lead.	*Note.*—At the conclusion of the wrought ornamental cut stone-work insert a full description of the manner in which it is to be secured together by plugs of slate, marble, stone, galvanised iron, or copper; copper cramps; and lead plugging, and running; bearing also in mind the channelling and lead running of water joints on the upper surfaces of cornices, &c.; the safe application of chain bars; the provision of sheet lead in the joints, and under caps and bases of columns, as well as between any other stones which, without lead, may have their meeting arrises crushed by vertical pressure.
Enrichments.	Particularly specify also the required accuracy, sharpness, &c., in the cutting of all enrichment, and the prior provision of satisfactory models (see No. 78); and expressly state that the work shall be cased over, and finally left perfect and clean at the conclusion of the whole.
Casing. Final perfection.	

MISCELLANEOUS STONE-WORK.

No. 127. Coping.
Cover the . . . . . . with . . . . . stone coping, of the sectional form and scantling shown by annexed sketch, throated under (one or both) edge(s) (qy. cramped with copper or galvanised iron?) and plugged at the joints with lead or slate. (Qy. chased to receive flushing?) (Qy. tooled, or rubbed?) No stone less than ′ ″ long.

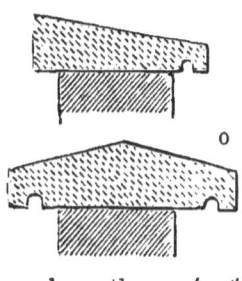

No. 128. Curbs.
Put round the . . . . . . a curb of . . . . stone, of the sectional form and scantling shown by annexed sketch (qy. how wrought?) (qy. cramped?), plugged at joints with lead or slate (and properly mortised for iron railing). No stone less than ′ ″ in length. (See Plumber, No. 364.)

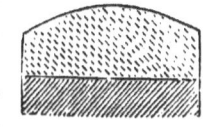

No. 129. Back hearths.
Put to fire-places proper back hearths of . . . . . stone   inches thick.

Front ditto.
Put to the fire-places of . . . . front slabs of (qy. slate?—Portland?—marble?) (rubbed or polished, as the case may be) not less than   inches thick; and   inches longer than their respective fire-openings. The same to be   inches wide.

No. 130. Chimney-pieces.
Provide and fix to the fire opening of . . . . . . a (slate) chimney-piece, valued at £   ; to that of . . . . . . a Portland ditto, value £   ; and to that of . . . . a marble ditto, value £   ; the Proprietor being at liberty to purchase all or any of these himself; the Contractor keeping distinct the allowance he has made for carriage and fixing.

[Devonshire supplies very beautiful, cheap,

and varied marbles for these purposes. Enamelled slate, or "Marezzo" marble may be used as substitutes.]

No. 131.
Paving, common.

Pave the . . . . . . with $\left\{\begin{array}{l}\text{Yorkshire}\\ \text{Purbeck}\\ \text{slate}\\ \text{limestone}\end{array}\right\}$ paving, not less than inches thick, and no stone less than feet superficial; the same to be well bedded on a good bottom (of dry rubbish) and jointed in mortar.

No. 132.
Better paving.

Pave the . . . . . . with $\left\{\begin{array}{l}\text{Yorkshire}\\ \text{Purbeck}\\ \text{limestone}\\ \text{slate}\end{array}\right\}$ paving, (rubbed or tooled) surface; and rubbed joints not less than inches thick, and no stone less than feet square; well bedded on a well-rammed bottom, and close-jointed in $\left\{\begin{array}{l}\text{mortar.}\\ \text{cement.}\end{array}\right.$

No. 133.
Superior paving.

Pave the . . . . with Portland stone, inches thick, surface and joints rubbed fine; laid (square? or diagonally?) in stones not less than ' " square, (or, as shown on drawings,) with cement under the joints, on a course of brick-flat, as sketch; the bricks being well flush-bedded in dry and well-rammed rubbish. (Qy. if any *marble* introduced with the stone?)

No. 134.
Marble paving.

Lay the . . . . . . with a paving, formed of the different marbles, and of the size and pattern shown and described on drawing; the whole to be executed with the finest possible joint, and geometrical exactness, and to be left thoroughly and uniformly polished. (N.B. If the marble be valuable, state the minimum thickness it may have as a *veneer* upon Yorkshire or slate stone.) The paving to be not less than inches thick,

PAVING, A BATH.

and the joints laid in cement, on courses of brick flat, firmly flush-bedded in dry well-rammed rubbish.

No. 135. Encaustic tesselated or Mosaic paving.

Lay the . . . . with the encaustic tile paving of . . . &c., allowing the sum of £ for the same, or at the prime cost of . . . per foot super. [Paving for ornamental purposes is now manufactured with marble tesseræ laid in concrete.]

No. 136. Stone fittings to cellar.

Form wine-bins in cellar with { slate Yorkshire &c. } slabs inches thick, on half-brick piers, as sketch, having neatly wrought edge, each slab the full length and depth of bin; and provide

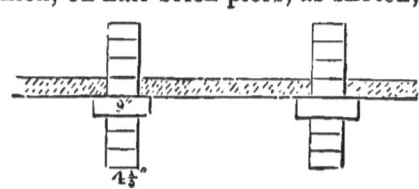

Larder. Dairy.

and fix also neatly wrought shelves of similar stone in the larder and dairy, cutting water channel, as described on drawing. [Or iron racks.]

No. 137. Trough, or sink.

Put in the scullery a neatly cut sink of . . . . stone, having a clear hollow of by and inches deep, with hole for waste water-pipe. [Doulton's stoneware sinks are preferable.]

No. 138. A bath.

Put in the . . . . . . a bath, formed of slabs of slate, grooved into each other, and bolted with iron, as sketch, and of the clear internal dimensions thereon shown; the

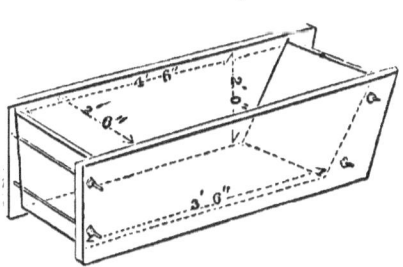

same to be internally lined with white glazed tiles bedded in cement; the top edges capped with (mahogany or marble) capping, and the exterior painted in imitation of white-veined marble. Form all necessary holes for supply and waste pipes.

No. 139. State any other stone or marble fittings; as slabs in halls, or passages; washing basins of marble, &c. [Ransome's Silicious Stone or other artificial concrete, as the Victoria Co.'s, are good substitutes for stone for chimney-pieces, truses, balustrades, and other useful and ornamental accessories.]

No. 140. STABLES, Miscellaneous Stone-work in—Sills and Steps.—Plinths to stall-posts, 8 inches square at top, 12 inches at base, neatly wrought 6 inches out of the ground, and 12 inches buried. Open surface gutter along front of stalls out of stone (8″ × 6″), and in lengths of not less than   feet.

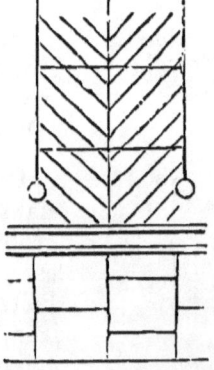

Sink-stones and gratings over drains.
Pavement. Pitch pebble-paving in sand.
    Do.   of dressed refuse stone, no stone less than 8″× 6″× 6″, close bedded in sand.

    Do.   of flat . . . . . . paving, no stone less than . . . . Those in stalls channelled to carry off wet; the rest rough tooled, and the whole well bedded and jointed on mortar. [Hard-burnt bricks or clinkers, purposely made for stable paving, are to be preferred to stone.]

Corn chest in loft, of slate or . . . . . . slabs, grooved into and bolted to one another, including a bottom ′ ″ by ′ ″, and sides and end   feet high.—A . . . . . . chimney-piece in saddle-room. (Qy. saddle-room paved?)

No. 141. COACH-HOUSES, Miscellaneous Stone-work in—Plinths of . . . . . stone to the coach door or story-posts, or piers, of the size and form shown and figured in sketch.

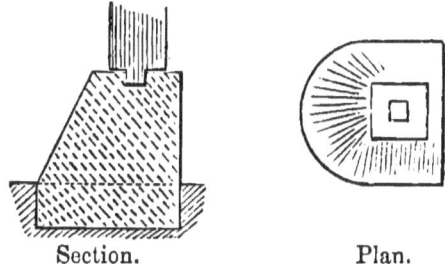

Section. Plan.

(If stone piers : state whether *wholly of wrought stone?*—in *one?* or how *many* stones ? or whether *hinge stones only* are required to be built into brick-work or rubble ?)

Stone to receive bolts of meeting doors.

Curb stone under doors from plinth to plinth. (Qy. whether *pebble paved?* paved with *dressed refuse?* or *flat-paved?* Guard stones to keep the wheels of different carriages apart, and to stop them at back.) (If there be a story of masonry above the coach-house door-openings, there will of course be *piers* instead of wood posts, and, instead of a wood bressummer, there will be lintels of stones, as *a, b, c, d, e,* jointed

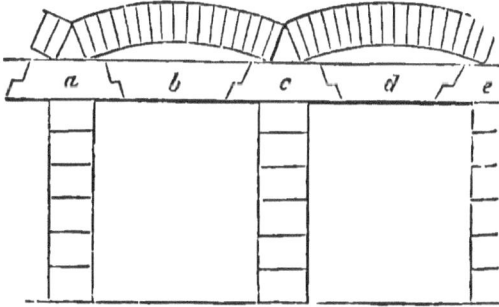

as drawings, and of the scantling thereon figured ; with *relieving arches* provided under the head of ' Bricklayer ' or ' Rubble Mason '). Describe any rebating there may be in the stone-work, and how the wrought stone-work is to be finished on the face—cramps, plugs, lead running, &c. Steps,

STABLE AND OTHER YARDS.

No. 142. Pitch-paving in sand, properly laid to a current, with sink and gratings.—Coping stones to dung-pit.—Coping to walls and boundary walls.—Stone caps to gate piers.—Gate piers, either partly, or wholly, of wrought stone.—Stones for hinges, bolts, &c.—Plinths to posts of sheds.—Curbs from plinth to plinth, and under gates.—Stone drinking-troughs.—Curb and cover stone to man-hole of tank.—Pebble or flat paving to cow-houses, piggeries. — Feeding-troughs to ditto.—Plinths to posts, and open stone gutters, &c., to cow-house.—Coping to outer pigsty, &c., &c.—Hinge stones to pigsty doors.—Curb stone under ditto.—Steps and sills to cow-houses, and other out-buildings having doors and windows.

SLATING.

No. 143. Slating, common. Cover the roofs with good scantle slate, on sound heart of oak or fir single or double laths, and oak pins; no slate to be less than " by ", and the whole to be well plastered against the pin with lime and hair mortar. The lap of upper slates over the lower to be not less than 2 inches. Properly cut double rag hips and eaves, and cut heading course.

No. 144. Better slating. Cover the roofs with $\left\{ \begin{array}{c} \text{large} \\ \text{small} \end{array} \right\}$ lady slates $\left\{ \begin{array}{c} 16'' \times 8'' \\ 14'' \times 7'' \end{array} \right\}$ nailed with cast-iron nails (boiled oil) to battens $2'' \times \frac{1}{2}''$, and well plastered underneath with lime and hair mortar. The lap of upper over lowest slate to be not less than $2\frac{1}{2}$ inches. Properly cut double rag valleys, hips, eaves, and heading course

SLATING. 255

No. 145.
Improved slating.

Cover the roofs with $\left\{\begin{array}{l}\text{countess, viscountess,}\\ \text{or large lady slates,}\end{array}\right\}$
" by ", nailed with cast-iron nails (boiled in linseed oil) to battens 2" by ¼", and pointed outside with putty of whiting, oil, and sand. The lap of upper over lowest slate to be not less than 3 inches.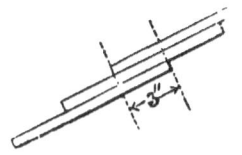
Properly cut hips, eaves, and heading course.

No. 146.
Superior slating

Cover the roofs with $\left\{\begin{array}{l}\text{queen}\\ \text{princess}\\ \text{rag}\\ \text{duchess}\\ \text{marchioness}\\ \text{countess}\\ \text{viscountess}\end{array}\right\}$ slates,*

"× ", nailed with copper nails to battens "× ". [Mixed metal or zinc nails are now generally used.] No slate to have less than a lap of $\left\{\begin{array}{c}4\\ 3½\end{array}\right\}$ inches over the lowest slate beneath it;

---

* Queens, 3' × 2', or 27" to 36" by irregular widths. 1 ton will cover about 2½ squares.

Ladies, 16" by 8". 1200 will cover 5 squares: weight 1¼ ton.

Princesses, 26 inches long, varying in width from 13 to 20 inches, averaging not less than 16 inches wide, weighing about 5 ton 3 cwt. per 1200, and covering about 16 squares.

Ditto, 28 inches long, varying in width from 14 to 21 inches, averaging not less than 17 inches, weighing about 6 tons per 1200, and covering about 18 squares.

Ditto, 30 inches long, different widths, from 15 to 22 inches, averaging not less than 18 inches, weighing about 7 tons per 1200, and covering about 21 squares.

Rags of large size, 17 dozen weighing a ton, and covering 2¼ squares.

Duchesses, 24 by 12 inches, weighing about 3 ton 7 cwt. per 1200, and covering about 11 squares.

Marchionesses, 22 by 11 inches, weighing 2 tons 14 cwt. per 1200, and covering about 9 squares.

Countesses, 20 by 10 inches, weighing about 2 tons 3 cwt. per 1200, and covering about 7¾ squares.

Viscountesses, 18 by 9 inches, weighing about 1 ton 13 cwt. per 1200, and covering about 6 squares.

and every slate to have at least two nails. Eaves, hips, and heading courses to be formed of cut slates, so that their bond may be uniform with all the rest.  All the horizontal over-lays to be well bedded (1½ inch up, from the edge) in (the stucco paint cement of Johns and Co., Coxside, Plymouth) or other cement. The raking or vertical meeting edge joints to be laid on a bedding of the same cement 3 inches wide.

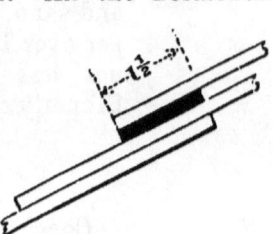

No. 147.   If the skeleton roof be of iron, the last description will serve, substituting wrought-iron laths, "strong copper wire" for hanging the slates, instead of "nails."

> [*Note.*—Iron roofs may be covered, 1st, by slating on wrought-iron laths; 2nd, corrugated or plain sheet-iron upon same; 3rd, by cast-iron galvanised plates fixed to rafters; 4th, plain sheet-iron on boarding.]

No. 148.   If the roof be over a circular building, state that the slates are to be cut to radiating joints from apex to eaves.

No. 149. Best slating for very flat-pitched roofs.   Cover the roof of . . . . . . with imperial slates not less than 2 feet 6 inches by 2 feet

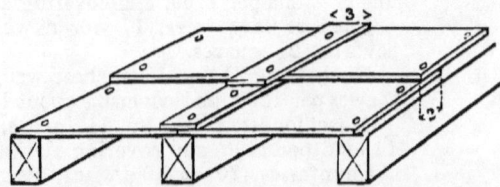

each, and full   inch thick: uniformly laid, with their ends meeting in a close joint along the upper surface of each rafter; each superior course to lap over the course below at least 2 inches; and the vertical or meeting joints to be covered with imperial slate slips not less than 3 inches wide. The over-lap of slates, and the slate slips, to be well bedded in (the stucco paint cement of Johns and Co., Coxside, Plymouth). Each slate screwed to the rafters with two 1½" screws, and two 2" screws to each slip. All visible edges of the slates and the slips to be sawn or rubbed to a perfect smoothness; and make uniformly close the cement pointing at the finish of the whole [or slate slabs, tongued and covered at joints by fillets or rolls].

No. 150. Outside pointing.
Where the slates are not *laid* in cement, they may be externally pointed, after laying, as follows: The over-laps and meeting joints throughout, to be made close with Johns's patent cement and sand, worked in with a stump brush, and the whole coloured as slates.

No. 151. Slate ridges, common.
Hips? (and) Ridges } to be covered with imperial slate slips, inches wide, inches thick, and in lengths of not less than ' ", securely screwed to rafters; close stopped at all meeting joints, and bedded on the slates 1½ inch up from the bottom edge with (Johns's patent paint) cement.

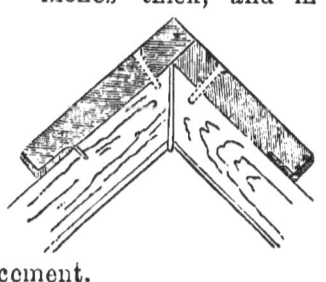

No. 152. Slate ridges, superior.
Hips? (and) Ridges } covered with imperial slate saddle-cut capping, of the size and sectional form figured and shown in sketch, and in lengths of ' "; securely screwed to rafters with screws, and jointed and laid in cement, same

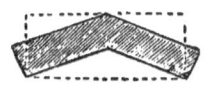

as the rest of the slating [or specify Robinson's patent ridge and hip roll bedded in oil cement and screwed].

No. 153. Filleting. Fillet the slating, wherever requisite, against the { brick-work / masonry } with (Johns's patent paint cement, mixed with equal parts of sand).

No. 154. Queen slating, of various sizes. Cover the roofs with . . . . The slates to be inches wide. Their length to commence (say 36 inches) long, at the gutters, and to diminish gradually to (say 30 inches) at the ridges; the same *bond* being observed throughout.

No. 155. Final clause. Examine, and perfectly make good, the whole of the slating, at the close of the works.
[*Note.*—Slate cisterns, shelves, screeds for hollow walls, &c., &c., should be specified here.]

TILING.

No. 156. Tiling, plain. Cover the roofs with good plain tiles on double heart laths, laid to a proper gauge { cement, lime and hair mortar, } in each plain tile secured by an oak peg.

No. 157. Ridge and hip tiles. The ridges (and hips?) to be covered with proper ridge (and hip?) tiles, secured by T nails dipped in pitch, and hip-hooks also pitched, and set in { cement. / lime and hair mortar.

No. 158. Pantiling. Cover the . . . . with the best sound pantiling, laid to a proper gauge, on pantile laths, and effectually pointed on the inside with lime and hair mortar. (*Note.*—If for the roof of a brewery, or other building, requiring ventila-

PLASTERING. 259

tion, or escape for steam, &c., the pantile *laid dry* is excellent.) Ridges and hips, as No. 157. [Ribbed Italian tiles, or those of the Broomhall Company, are recommended for appearance.]

No. 159.
Final clause.
The whole of the tiling to be left perfect at the close of the works, and no mortar to show externally to the disfigurement of the surface.

PLASTER AND CEMENT-WORK.

No. 160.
Johns & Co.'s patent paint cement, inside work.
Cover the partitions and battened walls with one coat of the paint cement, mixed with very fine sharp and clean sand, on lath and first coat of common lime and hair plaster. Cover the rubble walls with the said paint cement, on a render of common plaster, as aforesaid. Cover the brick walls with simply one coat of the said cement. The cement to be worked to a $\left\{ \begin{array}{c} \text{fair} \\ \text{or fine} \end{array} \right\}$ surface with a $\left\{ \begin{array}{c} \text{wood} \\ \text{or steel} \end{array} \right\}$ float; and the whole to be carefully applied according to the printed instructions of the Patentees.

No. 161.
Parian and Keene's cement.
Parian or Keene's cement may supersede wooden angle beads, &c., &c.

No. 162.
Commonest internal plastering.
Lath, lay, and set the ceilings and partitions of . . . . . ., and render, set, the walls of . . . .

No. 163.
Common 3-coat work for ceilings and papering.
Lath, lay, float, and set the ceilings and partitions of . . . . . ., and the battening against walls; and render, float, and set the unbattened walls.

No. 164.
3-coat work for painting or colour.
Lath, lay, float, and rough stucco the partitions and battened walls; and render, float, and rough stucco the unbattened walls of . . . . (Qy. whether jointed to imitate ashlar?)

No. 165. Best 3-coat work for paint or paper, &c.	Lath, lay, float, and finish with trowelled stucco, the partitions and battened walls; and render, float, and finish with trowelled stucco, the unbattened walls of the . . . . (Qy. if jointed?)
No. 166.	To whiten all the ceilings.
No. 167.	Colour the walls and partitions of . . . . a . . . . . . colour.
No. 168.	Run all beads, quirks, &c., to angles of arched soffits, and where else required. To properly plaster all sides, backs, soffits, &c., of window or other recesses, arches, ceilings under stairs, and other parts not cased with joinery, so that they may finish in conformity with the adjoining plastering.
No. 169. Cornices and enrichments.	Run, all round the ceilings of the various rooms, the cornices, and execute the various enrichments, as shown and described on the drawing of "Plasterer's Details," sheet No. . Models of all enrichments to be first made, (qy. at the expense of the Contractor?) and casts therefrom finally approved by, and deposited with, the Architect, before the enriched work be commenced. [Give girth or width of cornices.]
No. 170. Cement skirting.	Run round the floors of . . . . a skirting of Portland (Keene's patent?—Roman?—Johns and Co.'s?) cement, of the size and sectional form shown by annexed figure.
No. 171. Scagliola, &c.	Execute, in the best Scagliola composition, the shafts of the columns, pilasters, &c., in the . . . . ; the Scagliolist engaging to provide the wood firring and cradling necessary to receive his work. The shafts to be in imitation of

{ verde antique ; \
  sienna ; \
  jasper ; } the caps and bases to be executed in Keene's patent cement, (qy. whether the entablature, or any part of it, is to be Scagliola, Parian, or Keene's cement ?) and the whole brought to the utmost polish, and left perfect.

No. 172.
Cement, outside brickwork.
Cover the whole of the external surface of the brick walls with one coat of the patent stucco paint cement, mixed with sharp clean sand, and applied according to the printed instructions of the Patentees, [or with Portland cement in the proportion of one of cement and three of sand.]

Or,

No. 173.
Common, on rubble.
Cover the whole of the external surface of the rubble walling with a render of common lime and hair, and one coat of the patent stucco, &c., &c. (See No. 172.)

Or,

No. 174.
Superior, on rubble.
Cover the whole of the external surface of the rubble walling with a first coat of the patent stucco paint cement, mixed with very coarse sand ; and a second coat of ditto mixed with a finer sand. (Qy. jointed ?)

No. 175.
Rough cast, on rubble.
Cover the external walls of the . . . . . . with a render and float of common lime and hair, slap-dashed with a rough-cast of fine clean-washed gravel and lime water. (Coloured ?)

No. 176.
2 coats, common, and 1 Aberthaw, on rubble.
Cover the external walls, where not otherwise covered, of the . . . . . . , with a render and float of common lime and hair, and a stucco of Portland or Aberthaw lime and fine sharp clean sand. (Qy. jointed ?) (Coloured ?)

No. 177.
1 coat, common, and 2 Aberthaw, on rubble.
Cover the external walls, where not otherwise covered, of the . . . . . , with a render of common lime and hair, a float of Aberthaw lime, &c., and a stucco of Aberthaw or Portland, jointed to imitate ashlar.

No. 178.
Portland, on brick.

Cover the external *brick* walls, where not otherwise covered, with a float of Portland, &c., and a stucco of the same, jointed to imitate ashlar.

No. 179.

Run, in properly prepared $\begin{Bmatrix} \text{Aberthaw lime} \\ \text{cement,} \\ \text{Portland cement,} \end{Bmatrix}$ all parts of the external work hereinafter described, viz.—

    a. The moulded cappings and plinths of chimney shafts.
    b. The top front and inside (to flashing) of parapets.
    c. The rail, balusters, and plinth of balustrade.
    d. The entire cornice, including the top surface thereof.
    e. The mouldings, enrichments, &c., &c., of the frieze and architrave.
    f. The strings, edges of rustics, channels of rusticated parts.
    g. The architraves of doors and windows of the ...... fronts, including the whole girth from the back of moulding at A, to the wood frame at B.

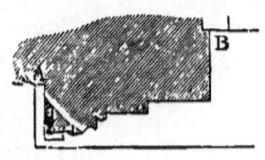

    h. The moulded work of the entablatures, pediments, of doors and windows.
    i. The cornices, trusses, &c., to doors and windows.
    j. The pilasters, columns, bases, and capitals, of doors and windows.
    k. The sills of windows.
    l. The top of main plinth, or the plinth entirely.
    m. The parapet and moulded work of portico, as cornice, architrave moulding, caps, bases, and (if fluted) shafts of columns, plinth, &c., and such other parts as are not to be executed in stone, or which cannot be as well finished in stucco.

The whole of the aforesaid Portland or Roman cement-work to be coloured in imitation of the other plastering.

[All the above features may be of terra-cotta, Ransome's patent stone, or other approved material, the architect selecting the designs, or they are to be modelled from his own designs.]

No. 180. All parapets, having lead flashings, to have a bed of cement right through them immediately above and touching the flashing.

[All brick and stonework above roofs to have damp-proof courses of slate in cement, sheet-lead, or asphalte to prevent damp drawing downwards.]

CARPENTERS' WORK.

No. 181. Carpenters' work, inclosures, &c.
Provide and fix all the timbers, boarding, &c., necessary to form the protective inclosures. (See No. 1.)

Construct also an office   feet by   feet clear, and   feet high to the springing of roof, for the Clerk of the Works, (see No. 1), the same to be formed of weather-boarding on stout framing, having a properly hung and glazed window, also a door in frame, with strong hinges and good lock; a drawing-desk   feet long by   feet deep, with drawer under; a stool; rail and pegs for cloak and hat; a corner cupboard with brass-knobbed latch. Floor properly boarded on joists.

No. 182. Shoring and old materials.
Provide and fix all required timber for shoring, &c. (See No. 2.)

No old timber to be used in the new works, unless permitted under the handwriting of the Architect.

No. 183.
Piling and planking.

Provide all the timber necessary for the piling and planking of the foundations, as described before. (See No. 10.)

No. 184.
Sundries.

Provide and fix all required scaffoldage, centering, turning pieces, beads, stops, fillets, tilting fillets, backings, blocks, cradlings, firrings, bearers, and all other minor articles of carpentry necessary to the perfect and efficient completion of the various works particularised under the heads of Carpenter, Joiner, Mason, Bricklayer, Slater, Plasterer. and Plumber.

No. 185.
Bond, &c.
Lintels.

Provide all necessary woodbricks and templates of sound Memel or red pine, with every required preparation for fixing grounds, battens, and joinery; also the various courses of bond timber and wall plates, described, shown, and figured on the drawings; also lintels of Memel or red pine over all square-headed window, door, or other openings, *within* the brick or stone arched soffits, it being clearly understood, with reference to *external* doors and windows, that no lintel shall appear *outside* the head of the wood frame. The said lintels to have a vertical depth of $1\frac{1}{4}$ inch for every foot of opening between the templates, and not to be longer than sufficient to cover the templates. One or more lintels, as required, to fill up for the thickness of the wall above; and the Carpenter to see that the relieving arches before described (see No. 56) are turned by the Mason. Templates to lintels not to exceed the scantling of 4"× 3".

No. 186.
Story-posts.
Provide and fix the story-post (or posts) as shown in drawings; the same to be of the soundest Memel fir, and of the full figured scantlings, with cast-iron boxed and tenoned caps and bases, as sketch, ⅜ths thick.

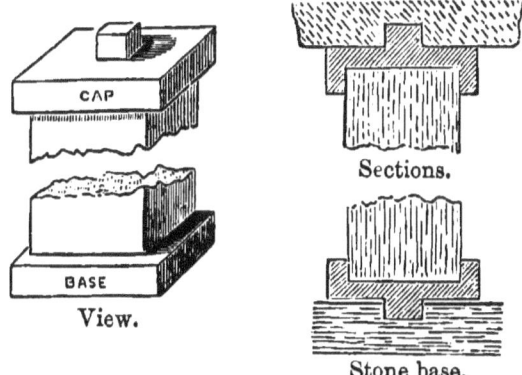

View.

Sections.

Stone base.

No. 187.
Bressummers.
Bressummers of the soundest Memel fir to extend over (here describe, whether from pier to pier, or over story-posts, or iron columns, or otherwise, wherever they have to be constructed contemporaneously and for the support of masonry) . . . .; the same to be of the full figured scantlings, and formed of single timber, halved, reversed, trussed with wrought-iron (king or queen) bolts, abutment ditto, struts, and straining piece, and bolted together with proper nuts, screws, &c., as shown by drawings [or a plate of wrought iron to be bolted between

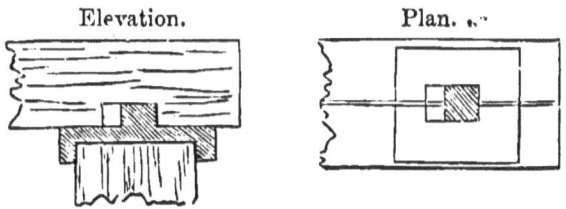

Elevation. Plan.

pieces]. The whole screwed up to a camber, and mortised for the tenons of the story-posts (or iron columns), taking care to leave the mortise free for a lateral thrust in the event of the camber settling again to a perfect horizontal.

N.B.—It is possible the bearing between the supports may be so small as to require no iron trussing. The Architect will here use his own discretion; as the *weight above*, or the *flooring bearing on* the bressummer *may* require his serious consideration. (See Part IV., Sect. 2, Strength of Materials.)

No. 188.
Quarter partitions.

See also "Construction" (Part IV. Sect. 2).

Sills of quarter partitions, resting on masonry, to be of sound old English oak, 4″ × 3″ (in very large buildings 6″ × 4″. The scantlings for the *usual* partition are here stated. When the partition exceeds 12 feet high, an increased size should be given). Heads and braces 4″ × 3″. Principal quarters, as door-posts, king or queen-posts, straining-pieces, &c., to be not less than 4″ × 4″. Common quarters 4″ × 2¼″, and 12 (16 or 18) inches from middle to middle (as the importance of the building, or of any particular part of it, may require). No quarter to have a length of more than feet, without horizontal stiffening pieces, as *a, a*. The whole, except the sills resting on walls as aforesaid, to be of (Memel fir on ground floors. They *may* be, if economy require it, of sound American red pine on the upper floors). All partitions hanging over voids to be truss-framed 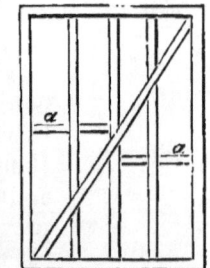 in the most careful manner with king or queen-posts, bolted with wrought iron to the sills or ties; struts and straining pieces to be properly framed into the same, and the whole rendered perfectly independent of the floor level with their sills.

N.B.—In particular cases sketches or drawings of these trussed partitions should be made; and it will be sometimes advis-

able to have the king and queen quarter of oak.
All required quarter partitioning to form closets, &c.

No. 189. Ground joists.
Ground joists to be of sound old English oak or Memel fir "× " and $\left\{\begin{array}{c}12\\14\\16\end{array}\right\}$ inches apart, on oak plates 4" × 3". [Space under ground floors should be concreted and ventilated.]

No. 190. Common joisting.
The rooms and passages, landings, &c., of (describing them with reference to the plans) . . . . . to be laid with $\left\{\begin{array}{c}\text{Memel fir}\\\text{red pine}\\\text{yellow fir}\end{array}\right\}$ joists; those of (such and such rooms) "× ", and inches mid to mid, &c., &c., &c. The whole to be properly framed into binders or trimmers, and to have a 6-inch hold in walls, bearing on plates 4"× 3". All trimmer joists to have an excess of 1 inch in thickness over the others.

No. 191. Binders and girders.
Here describe any binders or girders that are to be employed in connection with the common floor joists, as to landings in staircases, or over any other internal openings where the joists cannot rest on walls or partitions.

No. 192. Single-framed floors.
The floors of . . . . . . to be formed of Memel fir binders "× ", and not more than 6 feet apart, on oak templates 4"× 4", and having a hold of 9 inches on walls, with bridging joists thereon of $\left\{\begin{array}{c}\text{Memel fir}\\\text{red pine}\end{array}\right\}$ 6" × 2", and $\left\{\begin{array}{c}12\\14\end{array}\right\}$ inches apart, and ceiling joists thereunder 2½"× 1½", and 12 inches mid to mid.

No. 193. Double-framed floors.
The floors of . . . . . . to be formed of Meme fir girders "× ", not more than 10 feet apart, on oak templates 6"× 4", having a hold of 12 inches on walls, with binders 8"× 5", not more

than 6 feet apart, framed into girders, and resting on oak templates 4"×4", with a 9-inch hold on walls. (Bridging and ceiling joists as No. 192.) If the girder exceed 20 feet long it must be trussed, as described for Bressummers, No. 187.

No. 194. Floor trusses, &c.
In churches, theatres, public rooms, &c., where there are galleries and floors rising in steps to different levels, accurate drawings must be given of the main trussed framework; and the specification will therefore specially refer to these drawings, as thus :—

The rising floors of . . . . . . to be supported on trussed framework, as drawings. The trusses (in such or such positions,—or not more than   feet apart) of the full figured scantlings, put together in the most workmanlike manner, and with wrought-iron bolts, straps, nuts, screws, &c., as drawings. The king or queen timbers of sound oak; the remaining timbers of $\left\{\begin{array}{l}\text{Memel fir;}\\ \text{red pine;}\end{array}\right\}$ binders for floor and ceilings; bridging joists or firrings for different levels of floor, and ceiling joists; the whole as drawings. Breast-work to front of galleries trussframed, as drawings, of material corresponding with the floor trusses, with iron bolts and straps as shown, and of the full scantling figured.

No. 195. Cross straining and Sound Boarding.
All (single-joist unframed) floors to have a range of cross bonding of fir pieces 2 inches square (as sketch), closely butted and firmly nailed between the joists at parallel distances not exceeding 6 feet. (This will not be done in inferior buildings, nor where the joists are to be visible. See No. 209.) [Properly fill in upon rough boards on fillets a   inch layer of pugging to prevent passage of sound. (See sketch.) Pugging may be of chopped hay and lime mortar.]

ROOFS. 269

Sound boarding. The spaces between the joists of the floors of . . . . rooms to be fitted in with sound boarding on proper fillets; and the Carpenter to see that the space between the said boarding and the floor boards be filled in with proper pugging of sufficient thickness.

No. 196.
Ceiling battens. The joists of (such and such) floors to have ceiling battens $1\frac{1}{2}'' \times \frac{3}{4}''$, and 12 inches mid to mid, underneath them, to insure a good ceiling for the rooms below.

No. 197.
Flats. Here introduce a description of the girders, binders, bridging and ceiling joists,—or of the binders, bridging and ceiling joists,—or of the simple joists only,—which may be required to support the lantern and lead flats of staircases, or the flats over porticos, bay windows, or other parts of the building; either making explanatory sketches on the specification, or referring to detailed drawings whereon the scantlings are all figured.

No. 198.
Lanterns. Here introduce a description of the rough carpentry necessary to raise the sill of the lanterns above the flats; also of the joisting or rafters necessary to form the flat or roof *over* the lantern. Scantlings as drawings.

No. 199.
Roofs, Italian. The roof over the . . . . . . to be supported by trusses, as shown and figured on drawing No. . These trusses to be not more than $\begin{Bmatrix} 6 \text{ to} \\ 10 \end{Bmatrix}$ feet apart, with half trusses at the ends of corresponding form and scantling; hip rafters " by ", properly framed into a dragon piece of oak " by ", the said dragon piece dovetailed into an angle tie of Memel fir, feet long, and inches square. The tie-beams to have a hold of $\begin{Bmatrix} 18 \\ 12 \\ 9 \end{Bmatrix}$ inches on the walls, notched on oak templates, inches long and inches square. King

or queen posts of wrought-iron rods or oak with wrought-iron step straps or bolts (or both), to unite them with the tie-beam. Principals of { red pine / Memel fir } also united to tie-beam with wrought-iron bolts, straps (or both), and the remaining timbers of roof, viz. the (assistant principals?) struts, (straining beams?) (straining sills?) ridge-piece, ridge-roll, purlins, pole-plate, and common rafters, to be likewise of { red pine; / Memel fir; } the whole framed in the most workmanlike manner, and the tie-beam to be wedged and bolted up to a camber of   inches. Valley rafters   " by   ".

[*Note.*—Iron is frequently and economically combined with timber in large roof trusses; all suspension pieces as kings and queens being of iron rods, with cast-iron socket heads and shoes to receive the ends of struts, &c.]

Continue to describe the secondary and collar-beam, lean-to roofs, &c., in their turn. For curb roof, see No. 204. For projecting eaves, see No. 207.

No. 200. Roofs, Gothic.   *Note.*—In these, it is likely there will be no tie-beam, nor any angle tie or dragon pieces. Collar-beams, hammer-beams, brackets, springing pieces, &c., will supersede the tie-beam. The valley rafters may remain; but there will be no half trusses at the ends. An accurate drawing of the roof must be made, the reference to it in the specification being only general. (See No. 203.) For curb roof, see No. 204. [Curved and moulded ribs bolted or grooved to principals are often used.]

No. 201. Dormer doors and windows.   Provide, frame, and fix all the rough carpentry necessary to form the dormer doors and windows in the roofs, as shown on drawings.

No. 202.
Boarding and battening for lead and slates, and Felting.

Lay the roof with $\{\frac{3}{4} \text{ or inch}\}$ rough Memel boarding for slates. Inch gutter boarding on proper bearers to parapets, and boarding for valley gutters, &c.; 2-inch drips. (See No. 203.) [All roofs not ceiled underneath should have felt inserted under slates as a non-conductor of heat and cold. Pugging with chopped hay and plaster is as good.]

Or,

Lay the roof with Memel fir battens $2'' \times 1''$, for slates. Inch gutter boarding, &c., &c., as before.

N.B.—If the roof have projecting eaves, there will be no parapet gutter boarding required.

Lay the several flats, roof of lantern, &c., with $\{\frac{3}{4} \text{ or inch}\}$ rough boarding for lead, forming proper rolls for joints and drips.

No. 203.
Open or Gothic roof.

It may so happen in the case of an *Italian* roof, and it will most likely occur in that of a *Gothic* roof, that the timbers are to be left visible from below; and that the specification must therefore describe whether they are to be "wrought fair," "cut," "chamfered," "moulded," or "decorated" as drawings. In a church, chapel, or Gothic hall, for instance, which has no plastered ceiling, "the roof," or "so much of it as is visible," would be so described; and the slates, or lead covering, instead of being laid on "rough boarding" (as in No. 202), would be laid on "inch deal boarding with (rebated and beaded) (or ploughed, tongued, and beaded) joints, and wrought fair underside."

No. 204.
Curb roof.

Curb roof over the . . . . . as drawings. Raking side, or sides, of $\{\text{Memel, red pine,}\}$ formed by framed and braced quartering, as described for quarter partitions, No. 108 : the tie-beams, king or queen posts, principals, purlins, struts, rafters, &c., &c., together with the iron bolts or straps, necessary to the strength of the work, to be referred to "as shown on drawings," and the description of the Italian roof, No. 199, to be followed out as far as it is suitable ; also the suitable particulars in Nos. 201 and 202.

No. 205. Garrets. Red pine binders from (tie-beam to tie-beam) or (collar to collar) " × ", and ceiling joists " × "; not more than inches mid to mid. Fir ashlaring from the (raking quarters, or the rafters) to the floor joists, as section. Dormer doors and windows, as No. 201. Trap doors, No. 206.

No. 206. Ceiling floors. Red pine binders " × ", and not more than feet apart, framed into tie-beams ; and ceiling joists " × ", not more than inches mid to mid. Chase-mortised into the binders. Openings for trap doors into roof where marked on plans ; and a rough boarded foot-way to be formed from the trap to the dormer doors, windows, or ventilators in roof.

No. 207. Projecting eaves. Provide and fix all rough carpentry necessary to form the projecting eaves, the finishings of which will be described under the head of Joiner.

No. 208. Troughs, cisterns, &c. Construct rough red deal water-troughs for lead lining, to conduct from . . . . . , through the roof to the cistern, or cisterns, at . . . . . ; the said trough to be inches wide and inches deep in the clear. Construct also the cistern, or cisterns, of the sizes and in the situations figured and shown in drawings, fixing the same on proper bearers. Cover to cistern, if it be outside, and feet same.

No. 209. Joists, wrought fair. *Note.*—In some cases, as in cottages, stables, &c., &c., the upper joists of floors and lofts will be wrought fair with chamfered edges or beaded angles; also the girders, binders, &c. (See No. 203.)

No. 210. Sundry rough work. Provide and fix all required fir bearers and rough carpentry to the stair flights, &c., &c., &c. Here go over the plans very carefully, from the roof through every floor downwards, and make as particular allusion as you can to the numerous minor features which may be peculiar to the design under consideration, and which cannot be considered in this general outline. Thus, in roofs and ceilings, ventilating apertures may be required. Skylights are to be prepared for, and quartering for rough boarded linings. Preparations may be required for hanging pictures; for hanging lamps to ceilings; for completing certain forms, which can only be done partially with real masonry; for forming jambs, arches, inclosures, roofing to porticos, porches, sheds, covered ways, or projecting windows, &c., &c., &c.

No. 211. Battening. The internal surface of the several walls, marked by a yellow line on the plans, to be prepared for the Plasterer with Memel deal battens $2'' \times 1''$, and 12 inches mid to mid.

No. 212. Cradling and firring. Prepare for the arched, groined, or coved ceilings; for the cornices, beam-work, panelled ditto; for the entablatures, pilasters, and all other work that is to be finished by the Plasterer, with good and sufficient cradling and firring.

No. 213. Columns, &c., and lowered ceilings. Provide and fix also all required rough carpentry to form the cores of Scagliola or wood columns, with whatever framing or trussed work may be necessary above them; and all binders and joists that may be required to such ceilings as are lowered beneath the floor joists above.

No. 214. Final clause.	The whole of the aforesaid Carpenters' work to be executed with sound and well-seasoned timber, free from sap, shakes, and injurious knots, and to be framed together with workmanlike skill and accuracy. The scantlings to be full, *after the saw;* and the iron bolts, straps, trusses, screws, nuts, &c., employed in the roofs, partitions, and bressummers, to be of the best iron, well hammered and wrought. The ends of all tie-beams, girders, binders, and other important bearing timbers, to be left free from mortar on all sides, so that the air may circulate around; and all joists, and the plates on which they rest, to have their concealed ends well coated with coal tar. Wherever the fir timber is not described as of Memel or Baltic, it will be taken up as American or best Swedish.
No. 215.	N.B.—In the event of sliding doors being desired between rooms, the quarter partitioning must be prepared accordingly with double framing, leaving the required space between.

JOINERS' WORK.

No. 216. Ventilator to roof.	Here let the specification allude to any drawing there may be for ventilator, turret, or other piece of joinery rising on the roof, and all ventilating hoppers, trunks, &c.
No. 217. Sky-lights.	Prepare and fix sky-lights, as drawing, of 2½-inch Memel casement and bars in proper rebated and beaded frame, with all required means for rendering the same weather-tight.

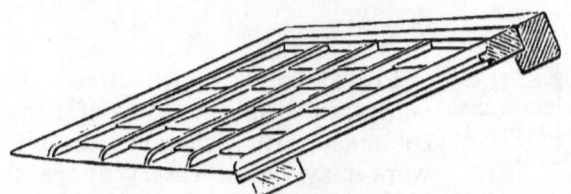

[The rafters for sash sky-lights to be grooved in rebates to let off water.]

JOINERS' WORK. 275

No. 218.
Ceiling, or dome inner lights.

Prepare and fix a 2-inch red deal neatly moulded light, in proper rebated and beaded frame, as sketch, in the ceiling floor under the sky-light, (qy. whether it is to open?) and inclose the space between the two lights with a neat deal boxing, leaving a door properly hung in one of the sides to allow of cleaning.

Or,

Prepare and fix a neat deal dome-light, in proper rebated and beaded frame, as sketch, in the ceiling floor, &c., &c., &c., as before.

No. 219.
Light and ventilation of water-closets.

It is well, when practicable, to light, and at the same time ventilate, water-closets by an adaptation of Nos. 217 and 218, lifting the sky-light on blocks so as to admit of air passing under the lower edge of the casement, and raising also the inner light for the same purpose. A dome-light would be best for a water-closet.

No. 220.
Dormer door.

Provide and hang in proper heads and jambs 2-inch Memel deal bead-butt and square dormer doors, as shown on drawings, with hingeing and fastenings complete.

No. 221.
Dormer windows.

Fit up, with $\begin{Bmatrix} 1\frac{1}{2} \text{ or} \\ 2\text{-inch} \end{Bmatrix}$ Memel $\begin{Bmatrix} \text{casements,} \\ \text{or sashes,} \end{Bmatrix}$ in $\begin{Bmatrix} \text{solid rebated} \\ \text{or deal cased} \end{Bmatrix}$ frames having oak sills (qy.

mullions ?) the dormer window openings, and fix window board; the whole as drawings.

No. 222. Trap door. Frame and fix an 1½-inch bead flush and square trap door, in proper rebated frame, for ascent into the roof where shown on plans, with hinges and fastenings.

No. 223. Gutter cornice to eaves, Italian, cantilever. Provide inch Memel deal cornice to eaves, also deal cantilevers, cut, moulded, and framed, at intervals of ' " apart, into an inch Memel deal fascia board, having moulding to correspond.

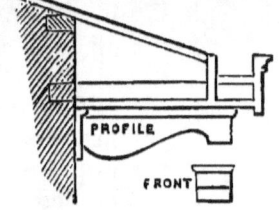

No. 224. Eaves cornice, Italian. Cornice to eaves to be formed of Memel deal, framed, glued, blocked, and moulded as drawing, with cut modillions at intervals of inches apart; and a gutter, of the clear dimensions shown in section, and having a falling bottom, to be formed behind the upper mouldings of the cornice.

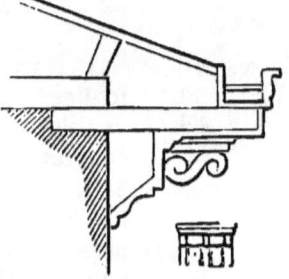

No. 225. Ditto, Gothic. Cornice to eaves to be formed of Memel deal, framed, glued, blocked, and moulded as drawing, with rain-water gutter, of the clear dimensions figured, and having a falling bottom formed in the upper part.

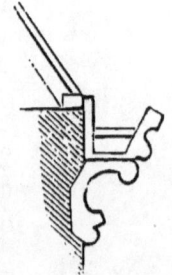

[Cast-iron gutters, plain and moulded, are now invariably used in preference to wood.]

BARGE-BOARDS.

No. 226.
Barge-boards.
The raking eaves of gables to be fitted with inch Memel deal, moulded and chamfered or beaded barge-boards, as drawing, to cover the ends of purlins.

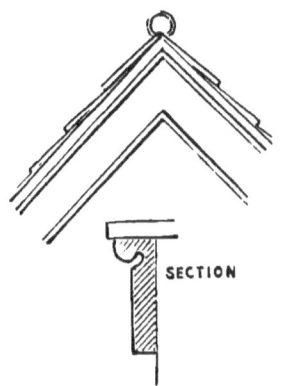

No. 227.
Barge-boards, Gothic.
The raking eaves of gables to be fitted with -inch $\left\{ \begin{array}{c} \text{oak} \\ \text{deal} \end{array} \right\}$ cut and moulded barge-boards, as drawing, the same to cover the ends of purlins.

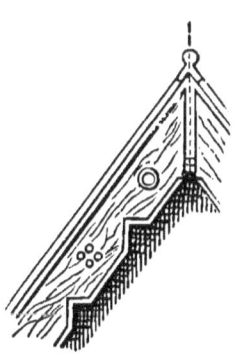

No. 228.
Lanterns.
Lantern-light over the . . . . to be formed of 2-inch Memel deal casements, in solid Memel fir rebated and beaded frame, with oak sill rebated, double sunk, weathered and throated, and fascia with moulding all round the top, from under the lead, to $1\frac{1}{4}$ inch over the joint between the head-piece and the casement. The whole to be wrought and moulded as shown by the draw-

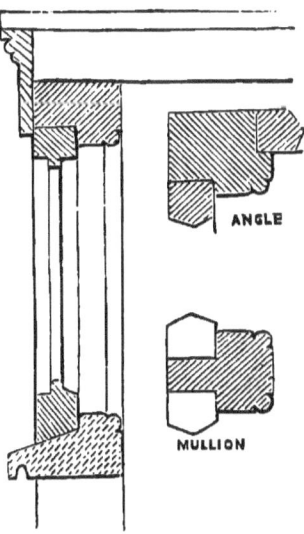

ings. (See No. 263.) (Qy. if any of the lights are to hang?)

[If it is intended to stain or simply varnish the inside woodwork, the framing of doors, skirtings, and other parts must be specified to be of selected deal, or pitch pine; the framing may be of yellow and the panels of red deal, all dead or loose knots and shakes being carefully avoided.]

No. 229.
Floors.
Boarding, common.

Lay the floors of . . . . . with inch . . . . deal folding; no board to exceed inches wide.

No. 230.
Ditto, better.

Lay the floors of . . . . . with $\left\{ \begin{array}{c} \text{inch or} \\ 1\frac{1}{4}\text{-inch} \end{array} \right\}$ deal, straight joint, face-nailed headings splayed; no board to exceed inches wide.

No. 231.
Ditto, superior.

Lay the floors of . . . . . with $\left\{ \begin{array}{c} \text{inch or} \\ 1\frac{1}{4}\text{-inch} \end{array} \right\}$ $\left\{ \begin{array}{c} \text{wainscot,} \\ \text{deal,} \\ \text{oak,} \end{array} \right\}$ straight joint, skew-nailed on one edge, and joints rebated:

or,

No. 232.
Ditto.

—and joints ploughed and tongued:

or,

No. 233.
Ditto, best.

—and joints dowelled with oak pins, at $\left\{ \begin{array}{c} 6 \text{ or} \\ 8 \end{array} \right\}$ inches apart:

(add to either of the foregoing three, viz. Nos. 231, 232, or 233,) headings ploughed and tongued, and no board to exceed inches wide; neatly mitred margins to hearths, and the boarding grooved for skirting.

State where floor boarding is to be wrought fair underside.

No. 234.
Very superior floor of deal and wainscot, or of wainscot wholly.

Lay the floors of . . . with $1\frac{1}{4}$-inch $\left\{ \begin{array}{c} \text{wainscot} \\ \text{or Memel} \end{array} \right\}$ best boarding, free from knots, and no board to exceed $\left\{ \begin{array}{c} 4 \text{ or} \\ 5 \end{array} \right\}$ inches wide, skew-nailed on one edge, dowelled with oak pins 6 inches apart; headings ploughed and tongued: and wainscot margin ——— feet wide (measuring from the skirting, which will be grooved into the same) all round the rooms, accurately mitred at all angles, with grain running parallel to the ends and sides of the room respectively.

No. 235. Parquetry and inlaid floors.
If the floors are to be laid in panellings and fancy patterns, an accurate drawing must be made and referred to.

[Parquetry floors may be laid of any design; or parquetry borders only. In this case specify manufacturer's name.]

No. 236. Skirtings, flush.
The rooms and passages . . . . . . to have ¾-inch deal beaded skirting, 4 inches high, and flush with plaster.

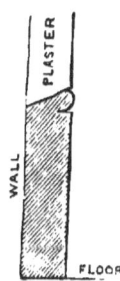

Or,

No. 237. Common skirting.
. . . . . ¾-inch deal hollow-moulded skirting, plugged to walls, 4 inches high.

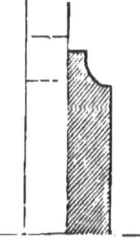

Or,

No. 238. Ditto.
. . . . ¾-inch deal skirting, 6 inches high; fillet and torus moulded, and plugged to walls.

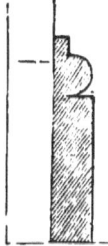

Or,

No. 239.
Skirtings and grounds.
.... ¾-inch deal skirting, with plinth 6 inches high, nailed to fillet and grooved grounds, and hollow and torus moulding above plinth.

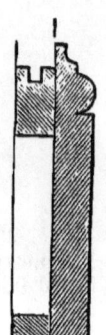

Or,

No. 240.
Ditto.
.... inch deal skirting; plinth 8 inches high, nailed to fillet and grooved grounds, and moulding above plinth, as drawing.

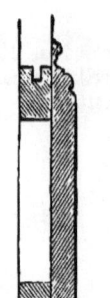

Or,

No. 241.
Ditto.
.... inch deal skirting; plinth 10 inches high, grooved into floor boarding, nailed to fillet and to 1¼ grooved grounds, and moulding above plinth, as drawing.

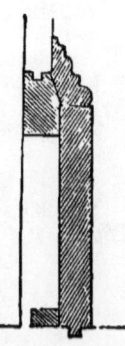

DOORS. 281

Or,

No. 242.
Skirtings.
..... 1¼-inch deal skirting; plinth inches high, with ¼-inch sinking to form double face, grooved into floor, nailed to fillet and to 1¼ grooved grounds, and moulding above plinth, as drawing.

[*Note.*—For kitchens, out-offices, &c., cement skirtings are best and cleanest.]

No. 243.
Doors.
Common
ledged.
The .... to have $\begin{Bmatrix} 1\frac{1}{4}\text{-inch} \\ \text{or inch} \end{Bmatrix}$ $\begin{Bmatrix} \text{oak} \\ \text{deal} \end{Bmatrix}$ door, formed with vertical ledges, rebated and beaded joints, nailed to three back braces, and hung with strong hook and twist hinges to solid rebated and beaded frame, 4″ × 3″, housed with iron shoe into step. Norfolk thumb-latch—wood stock lock.

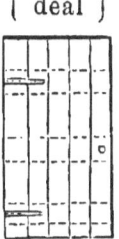

No. 244.
Ditto, framed
and braced.
The ...... to have door formed of 2-inch Memel deal stiles and top rail, filled in with inch deal battens, rebated and beaded joints, and backed with two horizontal (and, if necessary, two diagonal) braces framed into stiles, &c., the same hung with strong $\begin{Bmatrix} \text{hook and twist} \\ \text{cross garnet} \\ \text{wrought-iron} \end{Bmatrix}$ hinges to solid Memel fir frame, rebated and twice beaded, and housed with iron shoe into step. Norfolk thumb-latch to each door—fine plate 8-inch stock lock.

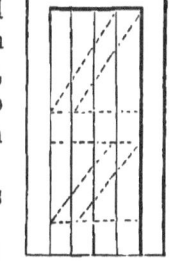

No. 245.
Coach-house
doors.
Coach-house folding doors will be the same as the last described; rebated in their meeting stiles, hung with strong wrought-iron hinges, and having bolts and swing bar.

No. 246.
Ditto,
common.
The ..... to have $\begin{Bmatrix} 1\text{-inch or} \\ 1\frac{1}{2}\text{-inch} \end{Bmatrix}$ yellow

No. 247.
Ditto, better.

deal four-panel bead butt and square (or bead *flush* and square) doors, hung with cast-iron butts to 1¼-inch jambs and heads; moulding on { one or / both } sides to cover the plaster joint; and iron rim brass-knobbed lock.

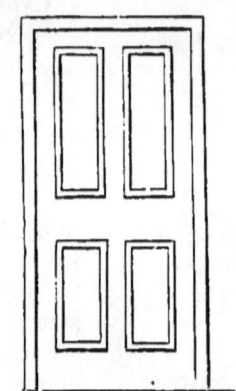

The . . . . . to have 1½-inch yellow deal four-panel, moulded (*both* sides; or one side, and the other square;) hung with iron butts in 1¼-inch single rebated jambs, with 2½-inch moulding on { one or / both } sides to cover plaster joint; and { mortise or / iron rim } brass-knobbed lock.

No. 248.
Ditto, superior.

The . . . . . to have doors formed of inch panels, in 2-inch stiles and rails, moulded both sides; (*one* side and square, in closets, &c.,) hung with two 3¼-inch best iron butts, in 1½-inch jambs double rebated and beaded, and { 6-inch / 5-inch } moulded architrave on framed and splayed grounds. Good mortise lock with ebony, china, or brass knobs to latch and bolt.

No. 249.
Doors, best.

The . . to have doors formed of 1½-inch panels, having sunk margins,

;

or sinking in centre

in 2-inch stiles and rails, the central style being double and beaded up the

middle, and the panels ogee and bead both sides. Door hung with best 4-inch lifting { brass or iron } butts, in 1½-inch double rebated jambs, and { 6-inch or 7-inch } moulded architrave on framed and splayed grounds. Best mortise lock with . . . . knobs to latch and bolt of approved manufacture.

No. 250.
Folding doors.

The opening between . . . . . rooms to have doors corresponding with the others in respect to their general character and mouldings, but hung folding, and having three panels in their height. Brass flush bolts top and bottom of one half, and the other furnished with mortise lock, &c., as other doors.

No. 251.
Sliding doors.

The opening between . . . . . rooms to have sliding doors, corresponding with the others in respect to their general character and mouldings, only that they will have an additional panel in height. The said doors to be hung by metal suspenders, to

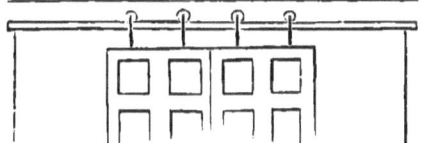

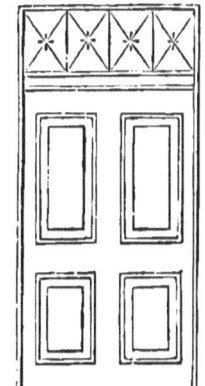

the axles of brass rollers, which will traverse a strong iron rail-rod. The whole to be executed in the best manner, and conformable to the drawings.

[Sometimes they are made to slide by friction rollers on metal rod let into floor.]

No. 252.
Outer doors.

The door-opening . . . . to be fitted with door formed of 1½-inch Memel deal ogee or ovolo bead and raised panels, three beads flush inside, in 2½-inch Memel stiles and rails, hung with two strong 4½-inch butts to solid Memel fir frame, 4½″ × 3″, rebated and twice beaded, housed with iron shoe into step, with transome, 4½″ × 3″, filled in with cast-iron rebated framework for glazing, as drawing: best 10-inch iron rim brass-knobbed lock, having draw-back latch, barrel, and chain. Wooden architrave on the inside, as drawing.

No. 253. ——— ..... (as the foregoing, with this variation) "hung folding, with two strong 4½-inch butts to each half,"—"one half to have a 2-feet barrel bolt at top, and 12-inch ditto at bottom; the other a best 10-inch iron rim lock, &c., &c."

No. 254.
Back doors.

The door-opening ..... to be fitted with door formed of 1½-inch Memel deal bead flush and square panels, in 2½-inch stiles and rails, hung with 4-inch butts in solid Memel fir frame, &c., &c., as No. 252.

No. 255. ——— .... as the foregoing, varying as No. 253 from No. 252.

No. 256.
Doors with side lights.

To either of the four foregoing doors, add

" side lights to the said door, as drawing, having

cast-iron rebated lattice-work for glazing, set in solid fir frame, corresponding with the rest, with $2\frac{1}{2}''$ panelled work below the lights."

No. 257. Door as the last, with *side*, but no *top* lights, and therefore no transome.

No. 258. If segment headed door, describe it, as drawing.

No. 259. If semicircular headed, describe it, as drawing.

No. 260. If pointed arched, describe it, as drawing.

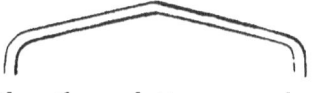

The top lights of the three latter may be termed fan-lights.

No. 261. If the doors be Gothic, the " ovolo and bead " or " ogee and bead " mouldings must be supplanted by " Gothic moulded, as drawings." It is impossible to specify for Gothic joinery, except as it regards the *substance* of the frames, transomes, stiles, and panels. Distinct drawings must be made for each particular case.

[Details of mouldings, &c. must be given in all cases.]

No. 262. Sundry doors. &c., Be careful that the general heading of "doors" include all dormer doors, trap-doors, blank doors, &c. Doors to casings of water-closet pipes, cupboard and dwarf doors, slid-

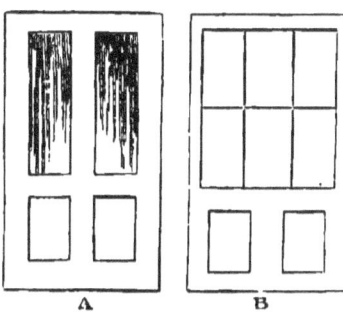

ing doors to buttery hatch, baized or cloths covered doors, with self-closing spring hinges; whether to open one or both ways? whether panel-glazed, as A, or casement-glazed, as B? and whether any of the inner doors are to be prepared for *borrowed lights* above them.

No. 263. Lantern-light.
Lantern-light over the . . . ., as drawings. Oak sill 6″ × 4″, rebated, sunk, weathered, throated, and grooved for lead. Angle standards of Memel or yellow deal 4″ × 4″, rebated and beaded; mullions 4″ × 3″, rebated and beaded; head ditto ditto. 2-inch Memel casements flush with the outside of frame (state if any are to open, and how?). Inch Memel fascia board and moulding for lead covering. Raking board grooved into sill to cover plaster cornice, &c.

Section.

Plan.
(See No. 228.)

No. 264. As above, with segmental, semicircular, or arched pointed heads, to the lights and frames, with spandril fillings, &c., as drawing.

No. 265. Windows, sash, simplest. The window-openings of . . . to have 2-inch or 2½-inch Memel deal sashes (double, or single, hung), with best flax, catgut, or other lines, iron weights, and iron axle pulleys, in proper deals cased frames, having oak double sunk sills 6" × 3". Good sash-*fastenings*. Inch angle bead to plastered jambs and *soffits;* inch rounded ledge to cap the plastered back, and skirting of room carried round the recess,

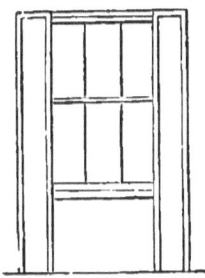
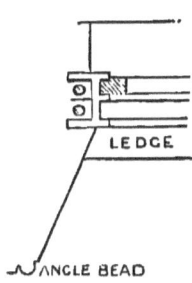

No. 266. Windows, sash, common.
... same as foregoing to the word "*soffits,*" continuing thus:—inch deal back, moulded and panelled to match doors, and skirting carried home to the same.

No. 267. Ditto, better.
..... same as No. 265 to the words "*sash-fastenings,*" continuing thus:—soffit, jambs, backs and elbows of inch deal, panelled and moulded to match doors, and moulding fixed on framed and beaded grounds to form architrave. [Brass axle pulleys instead of iron to be provided.]

No. 268.
... same as last with the addition of the margin M, for Venetian blind.

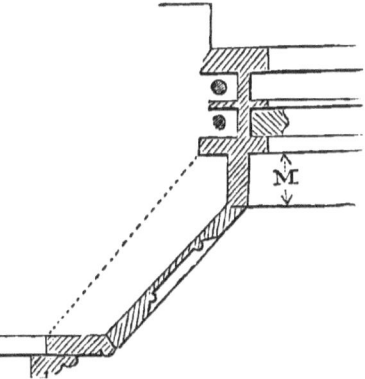

No. 269. Ditto, improved.
Same as No. 265 to the words "*sash fastenings,*" continuing thus:—soffit, back, elbows, and properly hung folding shutters; the whole panelled and

moulded to match doors, except the back flaps, which will be bead butt and *square*, to fall back against plastered jambs, in boxing formed by the grounds, and moulding fixed on the latter to form architraves as to doors. (Qy. if the addition No. 268 ?) See No. 271.

No. 270.
Windows, sash, best.

Same as last to the word "*square*," adding:— to fall back against proper bead butt and square back linings forming boxing with the grounds, and architraves complete as to doors. (Qy. if the addition No. 268 ?) See No. 271.

No. 271.

The shutters and back flaps to be properly hung with strong butt hinges; shutter latches, with furniture to match doors; and strong wrought-iron locking bar.

No. 272.

Same as No. 265 to the words "*sash-fastenings*," continuing: additional cased frame for lifting shutters to be hung as the sashes, and to descend into a proper deal casing, as drawing, having hinged ledge at top. The shutters bead butt and square, with brass-headed iron screw in brass screw-hole, to fasten them, as shown by sketch. The front of shutter casing panelled as doors, and architrave also to match.

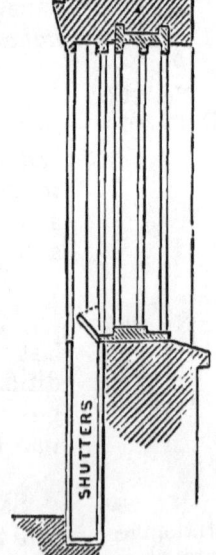

No. 273
Improved pulley stile.

[An improved kind of pulley stile, in which the weights may be taken out by a movable pocket-piece fitted with brass flush, ring, &c., and new lines introduced, or the sashes easily removed

WINDOW SASHES.

and reversed for cleaning, has been patented, known as "Gurman's patent."]

No. 274. Same at last, with the addition of soffit jambs and elbows, panelled as front of shutter casing.

No. 275. Three light windows. The window-opening of .. to be fitted with $\begin{Bmatrix} 2\text{-in. or} \\ 2\frac{1}{2}\text{-in.} \end{Bmatrix}$ Me-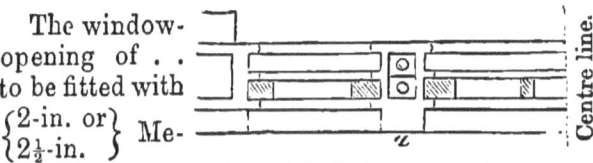
mel deal sashes, in triple-light cased frame, as drawings; the central sashes double hung with best lines, iron weights, and brass pulleys:— continuing to describe the rest of the inside joinery, as may be selected from Nos. 265 to 274 inclusive. (See No. 279.)

No. 276. Bow ditto. If a *bow* window, describe it such. (See No. 279.)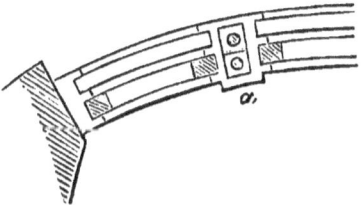

No. 277. Bay, ditto. If a *bay* window, say, "the openings of bay to be fitted with $\begin{Bmatrix} 2\text{-inch or} \\ 2\frac{1}{2}\text{-inch} \end{Bmatrix}$ Memel deal sashes, in proper cased frames," &c., stating whether the

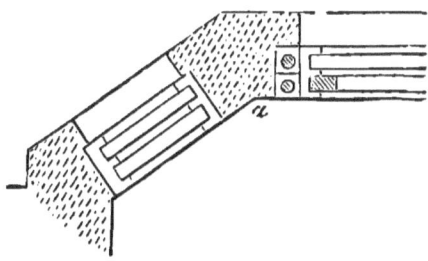

central sashes only, or the whole of the sashes, are to be double hung, &c., &c. (See No. 279.)

No. 278. Venetian window.	If a *Venetian* window, say, "the openings of Venetian window to be fitted," &c., &c., &c., as before. (See also No. 279.)

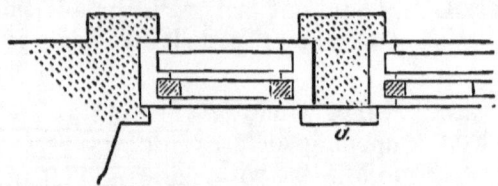

No. 279.	The windows, Nos. 275 to 278, may have "wooden casing pilasters" at *a, a, a*, &c., which must be described "as shown in drawings."
No. 280.	If the frames and sashes have "segment," "semicircular," or "pointed arched heads," describe "as drawings."
No. 281. Windows, casement.	Fit up window-openings in . . . . . with -inch deal casements, filled in with iron bars and lead (qy. diamond?) work for glazing; (certain of them to be) hung with strong butt hinges, in solid fir, wrought, rebated, and beaded frames, and oak sills, as drawing (state if there be *mullions* or *transomes*). Inch deal window-boards. Good fastenings to close, and approved means for holding open, the casements. (See No. 284.)

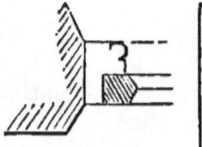
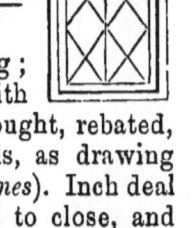

No. 282.	Same as last, omitting "filled in with iron bar and lead for glazing." (See No. 284.)
No. 283.	If inside shutters, soffits, backs, or elbows, select from No. 265 to 279. If outside shutters, make them conform with the doors described in Nos. 243 to 246, adding the required fastenings for securing them when open or shut.
No. 284.	If the windows be Gothic, the casements will be described as hung in solid frames (mullions) (transomes) and oak sills, moulded, as drawings, with their hinges, bolts (top and bottom), latches,

CASEMENTS. 291

and fastenings, stating whether pointed headed, &c. Soffits, shutters, backs, elbows, &c., as from No. 265 to 279.

No. 285. French casements.
The window-openings of . . . to be fitted with 2-inch or 2½-inch Memel moulded and rebated casements, having vertical (and transverse?) meeting stiles (and rails?) as drawings. The same to be hung in solid Memel fir, wrought, rebated, and beaded frames $\left\{ \begin{array}{c} 5'' \times 3'' \\ 4'' \times 3'' \end{array} \right\}$ in oak double sunk sills, with strong 4-inch butts: good brass-knobbed latches, and (brass bolts top and bottom of the opening casement) (or) patent rod-bolt to make the meeting casements, when closed, perfectly tight. For soffits, shutters, backs, elbows, pilasters, &c., &c., see Nos. 265 to 279.

[The stiles of all casements should be tongued to fit into grooves made in the frames, or *vice versâ*, meeting stiles should also have a hook rebate joint to prevent wet driving through, and galvanised iron sill bars inserted in groove with white lead.]

No. 286.
The windows of . . . . . . to be fitted with . . . . deal swing casements to hang on centres, as drawing, with apparatus for opening, closing, and fixing the same, in solid rebated and beaded frames and oak sills, as drawings.

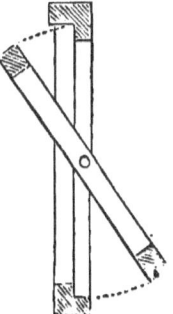

[Iron casements and frames with check rebated joints are preferable in many cases.]

o 2

No. 287.   *Note.*—Be careful that all common windows and doors, which have no fittings of joinery inside, be furnished with coin-beads, and that such windows have window-boards.

No. 288.   Fit up the . . . . . window-openings with neatly wrought luffer-boarding, (state whether *fixed*,—or if it is to be made to close or open at pleasure, by each board revolving on a wood pin at each end,) in solid Memel fir frame and oak sill. (If the boarding is to open and close, a rod and peg, as sketch, must be made to revolve on pins in the head-piece and sill, with a handle, or other means for turning it.)

No. 289.   Frame and fix a clock or bell-turret over the . . . . . . ; the same to be of Memel fir, and agreeable to the drawings and the descriptions thereon.

No. 290.
Stairs, common.
The . . . . staircase to be fitted with inch deal steps having rounded nosings, on ¾-inch risers, framed into skirting and outer strings, with neatly turned or square newels and inch-square balusters, two on each step, and neat 2¼-inch rounded hand-rail. (See Nos. 293 and 294.)

No. 291.
Stairs, better.
The . . . . staircase to be fitted with 1¼-inch deal steps, having moulded nosings, framed into skirting and outer string; the latter wrought, moulded, and capped, as drawing, with 1¼-inch balusters; wainscot, hand-rail, and newels, cut and moulded as pattern. (See Nos. 293 and 294.)

No. 292.
Stairs, best.
The . . . . staircase to be fitted with 1¼-inch deal steps, with moulded nosings along fronts and returns; inch deal risers; end-casing of steps cut as pattern, and fascia moulded as draw-

ing; handsome curtail step, {wainscot / mahogany} hand-rail, and turned {wainscot / mahogany} balusters, two to each step. (See Nos. 293 and 294.)

No. 293. A sufficiency of cast-iron balusters of corresponding pattern, to give stiffness to the hand-rail. [Cast-iron balusters may be used instead of wood.]

No. 294. The face of landing to have nosing and fascia corresponding with the steps and balusters also.

No. 295. The staircase of . . . to be fitted (as No. 291 or No. 292), adding, " and the soffit of stairs to be panelled with wood, as shown and described on drawings."

No. 296. The . . . . staircase to be fitted with (as No. 291 or No. 292), adding, " and the space (or a certain part of it) under stairs to be inclosed with wood-panelled spandril, moulded outside as drawings, and square on the inside."

No. 297. The staircase of . to be (as No. 291 or No. 292), excepting that " cast-iron ornamental balusters" will supersede the wooden ones.

No. 298. The . . . . . to be approached by a flight of inch {oak / deal} treads framed into 1½-inch string-bearers.

No. 299. Frame and fix at . . . . . and . . . . , &c.
Spandrels. {1-inch / 1½-inch / 2-inch} {wainscot / deal} panelled spandrils, moulded or square, so as to correspond with the doors or other joinery with which they are seen or connected. (State whether there are to be skirtings or cornices, and if any borrowed lights therein.)

No. 300.
Casings.
Specify whatever *casings* (plain, panelled, or beaded) there may be, not before mentioned, such as to beams, lintels, bressummers, storyposts.

No. 301.
Specify if any boarding (plain, grooved, tongued, and beaded, or otherwise) on battens against walls, or against quarter partitions; or whether any panelled casings against the same, as dados, or wainscoting; and whether base and surbase mouldings.

No. 302.
The columns and pilasters in the . . . . to be formed of 2-inch yellow deal staves, glued and blocked with turned yellow deal bases, and caps (if Tuscan or Doric). (If the caps are Corinthian or Ionic, they will be described as " carved according to the drawings.")

No. 303.
Moulded architrave, &c., frieze, and glued, blocked, and moulded cornice (describe whether enriched, or with modillions, dentils, &c.).

No. 304.
Describe all other carved work.

No. 305.
Papier-mâché work, &c.
Describe papier-mâché, carton-pierre, or gutta-percha work in connection with joinery. Where mouldings of panels, architraves, cornices, &c., &c., in the joinery, are to be *gilded*, they should always be enriched; and papier-mâché is the best way of preparing for it. See published Book of Ornaments, and Tariff of Prices.

No 306.
Fit up the water-closet with -inch $\begin{Bmatrix} \text{deal} \\ \text{mahgny} \\ \text{cedar} \end{Bmatrix}$ pierced seat, and properly framed $\begin{Bmatrix} \text{(brass-} \\ \text{(iron-} \end{Bmatrix}$ hinged) clamped flap of the same material, with moulded nosing to range with fixed sides, having boxed sinking for handle in one, and larger ditto ditto for paper in the other. Back and elbow beaded boards, 8 inches high, of same material

STABLES. 295

as flap, &c. The riser to be of . . . . . . panelled, and casing for pipes, with doors, &c., as plan.

No. 307. Privies fitted with deal riser, seat, and cover, and back and elbow boards.

No. 308. Sundry fittings—as glass-washing troughs in butler's pantry, dish-washing ditto in scullery, and all such presses, shelvings, cupboards, &c., &c., as are positively connected with the permanent building; wooden chimney-pieces; rails and pins in closets, passages, and halls, &c., &c., &c.; fittings in butler's pantry, housekeeper's room, china closets, store rooms; knife and shoe, drying and brushing rooms; wash-houses, laundries, brew-houses, cook's closets, housemaid's closets, larders; salting-rooms, dairies, scalding-rooms, still-room, &c., &c. [Specify wrought and dovetailed inch deal cistern case.]

STABLES.—Miscellaneous Joinery; as

No. 309. Stall-posts 6 inches square, chamfered, with iron shoes; head-piece to match; stall divisions of -inch vertical oak boarding, ploughed and tongued joints, grooved into bottom rail 4" × 3", and into rounded top rail 4" × 5".—Racks ⅔rds width of stall, formed of rounded oak bars 1⅛-inch diameter, in rounded top rail 5" × 4", and bottom ditto 3" × 3". Boxed oak mangers, occupying ⅓rd width of stall; standard-post properly fixed in ground, with halter-pulley therein: the backs of stalls boarded as high as the stall divisions, and boarding round all the walls beside, ranging with the stall boarding.—Lockers for corn, &c.—Traps for letting down hay.—Harness-room boarded top to bottom, with harness-pegs and saddle-trees, closets, &c.—Steps or ladders to loft.—Doors, windows, hay-loft doors, &c.—Ventilating trunks.—The loft-joists and boarding wrought fair where visible from below.—Trunk and apparatus for obtaining corn from the chest above. (See No. 313.)

No. 310. COACH-HOUSES.—Folding doors in jambs, heads, &c.—Story-posts.—Bressummer over . . . . door from saddle-room. (See No. 313.)

No. 311. Loose boxes boarded round as stables, with angular quadrant racks and mangers, of the same general character as to stalls. (See No. 313.)

No. 312. All required joinery to outhouses and coach gates, sheds, &c., cow-houses, piggeries, &c., &c. (See No. 313.)

No. 313. The doors and windows suited to stables, coach-houses, outhouses, &c., will be found against Nos. 243, 244, 245, 246, 281, 282, 286, 287, 288; stable clock turret, 289. [The best stable windows are of iron, pivot hung casements in small squares.]

No. 314. Final Clause. The whole of the aforesaid joinery to be executed with sound and well-seasoned timber, free from sap, shakes, and large or loose knots, and to be so early prepared that, after its fixing, it may remain secure from serious shrinkage. All obviously necessary or usually required ironmongery to be supplied and fixed by the Joiner, whether specified or not; and all hinges, locks, latches, bars, catches, bolts, &c., to be left perfect at the close of the works, easy in their action, and free from rust.

IRON AND METAL WORK.

No. 315. Window guard bars. The windows of . . . . (*i.e.* such as have no shutters, but where security by night is required) to be fortified by strong wrought-iron bars, firmly screwed into the sills and heads [or provide Moline's patent wrought-iron windows].

IRON AND METAL WORK. 297

No. 316.   The windows and .... of .... (dairies and larders, for instance) to be fitted with fly-wire or perforated zinc instead of glazing;
Or,
The windows of .... to have separate frames filled in with fly-wire, so that the glazed casements or sashes may open independent of them.

No. 317.
Metal lights.
Fix over the opening of .... a skylight of cast-iron, or $\left\{ \begin{array}{c} \text{zinc, or} \\ \text{copper,} \end{array} \right\}$ formed as shown by drawing, the bars being of the section as sketch.
(Rolled iron moulded ribs are now used.)

No. 318.
Chimney bars.
Put to the fire-places of .... wrought-iron chimney-bar $2\frac{1}{2}'' \times \frac{1}{2}''$, properly caulked at the ends.

No. 319.   Qy. bolts to prevent brick hearth trimmers from spreading.

No. 320.
Iron columns.
Provide and fix cast-iron columns to support the ....; the same to be of the best iron, cast hollow; the entire diameter being   inches, and the thickness of the iron being not less than $\frac{1}{8}$th of that entire diameter. Cast-iron plate as drawing on the top.

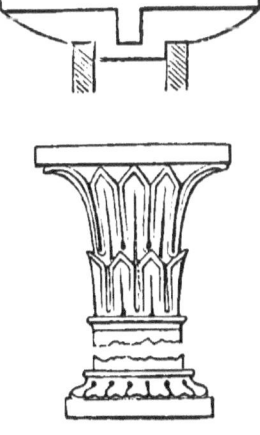
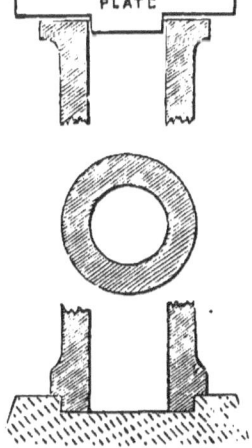

Qy. bracket head, as drawing?

Qy. caps and bases, plain or enriched, as drawing?

No. 321. Iron girders.
Provide and fix cast-iron girders over the . . . . ; the same to be of the best iron, of the form and full scantling (after shrinkage) shown in drawing?

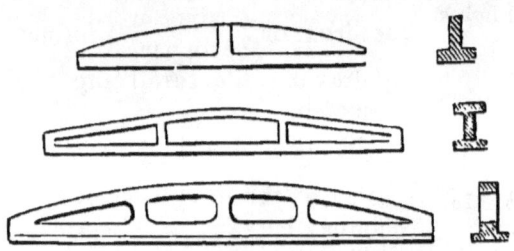

No. 322. Iron joists.
Provide and fix cast-iron joists (to receive either brick arches, stone floors, &c.?); the same to be of the best iron, and of the full scantling (after cooling) shown by section.

Or,

[Provide and fix rolled-iron joists of section shown.]

Or,

[Provide and fix wrought-iron rivetted plate girders of the scantlings and section given (here specify whether "box," or lattice, or I-shaped sections are required, also maker's name and required net strength. Angle irons, rivetting, stiffening pieces, &c., should be accurately shown and described).]

Or,

No. 323. Fire-proof floors.
[The fire-proof floors to be constructed as per drawings, and to have wrought, cast, or rolled iron girders (here describe particulars of construction, as distances apart, fixing, &c.), to receive brick arches or concrete.

ROOFS, GUTTERING. 299

Or,

Provide and fix (state here name of some particular system of fire-proof flooring or patent, as the "Dennett" arch; Phillips & Co's; Shaw & Co's, Hornblower's, &c.)]

[The girders to be tested to within two-thirds or one-half of their calculated breaking weight. (See "Iron Construction," Part IV.)]

No. 324.
Sundries.

Wrought-iron tie-bolts? cast-iron plates? chain bars.

No. 325.
Iron roofs.

Iron roofs can be only generally described as " of the best wrought and cast iron, and of the forms and full scantlings (after shrinkage) shown by the drawings," which cannot be too much detailed and described thereon. (See Bartholomew's Specifications, chap. 47.) [See section on Ironwork.]

No. 326.
Gutter cantilevers.

Provide and fix . . . . cast-iron cantilevers to match with the wooden ones of . . . . cornice; the same to be cast hollow to act as gutters in conveying the water from the gutter in front to the heads of water-pipes.

No. 327.
Guttering.

Provide and fix along the eaves of . . . . roofs a cast-iron gutter, as sketch, with proper bracket supports, &c.

Or,

[Provide and fix to eaves of roof cast-iron guttering of the section shown (or here specify the castings of some approved manufacturer, Macfarlane's, Smith & Co's., &c.). Gutters are made of various moulded forms; also stack pipes.]

Or,

Zinc shuting may be used.

No. 328.
Water-pipes.

Provide and fix the various cast-iron water-pipes shown on plans; those of . . . to be of inches clear bore, those of .... of inches ditto, and those of .... of inches ditto. All these pipes to have proper receiving heads, with roses or gratings to prevent the descent of leaves or rubbish, and proper shoes at the bottom to turn off the wet from walls. The inside of pipes to be painted three times before fixing.

No. 329.
Gratings.

Provide and fix { wrought- or cast- } iron gratings over the areas of ..... ; the same to open (single or folding) on centres, with means for securing the same when closed.

Or,

[Provide and fix patent prismatic pavement lights. (Specify the kind, such as Hyatt's.)]

No. 330. Qy. any fixed gratings?

No. 331. Provide and fix the various cast-iron gratings to air-holes for ventilating underground joists; also those required over cesspools, to the gutters or drains of .... ; also the several coal plates, with means for securing the latter inside, and iron stench traps.

No. 332.
Rails and balusters, &c.

Provide and fix the iron railing or palisading, wrought *plain*, or cast *ornamental*, as indicated in plans, and shown in detailed drawing; also the rail and balusters, wrought *plain*, or cast *ornamental*, to the areas, landings, balconies, steps, stairs, &c., &c.; the whole to be properly fixed, with screws, &c., in *wood*, or with lead running in *stone*-work, as the case may be.

No. 333.
Gates.

Provide and hang with proper wrought-iron hinges and centres (qy. revolving in brass cups?) the wickets or gates, single or folding, of wrought-iron *plain*, or of cast-iron *ornamental*, shown in drawings; and with the bolts, stops, latches, locks, &c., thereon described.

No. 334. Iron doors. — The opening into strong closet, plate closet, &c., to be fitted with an iron door and lock valued at £ .

No. 335. Wood and iron ditto. — A good door may be formed by sheet iron panels, screwed and rebated into wrought-iron stiles and rails, moulded in front and flush at back, the latter having fixed to them wooden stiles and rails to form panels like the front. The whole to be hung with strong 5-inch butts in iron rebated head and jambs; and brass lock on the iron side.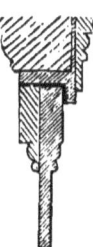

No. 336. Iron casements. — The windows of . . . . to be fitted with cast or wrought iron casements, as fully shown and detailed on drawings. (The drawings will show the frames or outer rims, the meeting bars, the common bars, the part to open, and the means of opening and closing.) Whether the windows open, or only ventilators *in* them?

No. 337. Sundries to windows. — If any particular bolts to make weather-tight the casements, or any metal slips to cover meeting joints of ditto, or to prevent water from passing over sill, describe them accurately: with any copper tubing that may be necessary to carry off wet or condensed moisture.

No. 338. Shutters. — In Banking-houses, business premises, or other buildings where fire-proof security is required, Bunnett's revolving shutters may be advantageously used, or Clark & Co's self-coiling ditto.

No. 339. — Provide and fix cast-iron mangers and racks, as per drawings, in stables.

Or,

No. 340. [Provide and fix complete (here state maker's name as) Cottam & Co's stable fittings, harness brackets, &c.

No. 341. Provide and fix verandah and covered way . . . of wrought-iron sash bars and cast-iron supports covered with lead or zinc (here describe weight of lead or number of gauge of zinc; also details and manufacturer's name).

No. 342. Sundries. Specify any special iron-work, as school desks, revolving shutters or hinges, ventilators, door springs, iron casements and frames, cellular iron bins, lightning conductors, &c.]

No. 343. General clause. Provide all the cast and wrought iron-work necessary to the completion of the Carpentry, according to common usage, whether herein specified or not, as spikes, nails, screws, holdfasts; also all cast and wrought iron or brasswork necessary to the doors, windows, shutters, lantern or sky-lights, and the joinery in general, as iron shoes to door-posts, hinges of the required varying description, locks ditto, latches ditto, bolts, bars, and chains ditto, and brass-knobbed handles.

Here proceed more minutely to describe the more important and particular Smith's work, as—

Wrought-iron abutment, king or queen bolts, struts, straining-bars, and coupling-bolts, with their washers, nuts, and screws, to trussed beams or girders; wrought-iron stirrups or straps and bolts, with their wedges, washers, nuts, and screws, to unite the king and queen posts and principals with the tie-beams of roofs.—Cast-iron shoes and cappings to wood story-posts.—Cast-iron box, sockets, or casings, to receive the ends of girders, binders, or tie-beams. Also any particular iron-work necessary to Stonemason's work. All castings to be clean, sound, free from air-flaws, &c., and, in all important cases, as with

GRATES, STOVES, ETC.

columns and bearing-beams, thoroughly *proved* before fixing. All wrought iron to be thoroughly welded and hammered.

No. 344.
Grates, stoves, ranges, &c.
Provide grates of approved construction for the various sitting and bed rooms, of the following prices respectively, viz. . . . . . . The grates in the . . . . rooms to have hobs.
 Provide stoves for the . . . . , of the following descriptions and prices respectively, viz. . . .
. . . . . Provide and superintend (including all carriage, men's time, and expenses) the fixing of a cooking apparatus and range for the kitchen, valued in themselves separately at £ . Fix smoke-jack of approved construction in kitchen flue. Qy. range or stove, or both, in back kitchen or scullery, valued at £ . Hot-plates, &c. ? Stove with oven, &c., valued at £ , in still-room. Scalding-stove in dairy scullery, valued at £ . Stoves and coppers in back kitchen or scullery, wash-house and bake-house, valued at £ . Ironing stove in laundry, valued at £ . Arnott's stoves in . . . . and harness-room, valued at £ . Qy. coppers and stoves in boiling-houses, brew-houses, &c., &c.

[Grates with fire lump backs and sides, warm-air chambers at back with inlets into the room either in the jambs or top of grate, are now generally adopted in good houses. Such are Captain Galton's stoves manufactured by Yates, Haywood & Co., of Upper Thames Street; Steer & Co.'s patent grate ; and the "Manchester grate" manufactured by Shillito and Shorland, Duke Street, Westminster, &c.]

No. 345.
Bell hanging.
Hang, on a proper board, painted and numbered, ... bells of varying tones, having springs and pendulums ; the same to communicate, by means of copper wire passing through $\left\{\begin{array}{c}\text{tinned or}\\ \text{copper}\end{array}\right\}$ tubing, (concealed in plastering,) with . . . . pulls, to be fixed where indicated on plans. The wires to be collected in the roof, and

to be attached with the utmost care to a sufficiency of cranks and coil-springs. The pulls to be of the best suitable kinds, with knobs or lever-pulls, as here described. The pulls in . . to be of . . . . ; those in . . . . to be, &c., &c.
[Or,
Provide and fix electric bells of approved manufacture.

Provide and fix all necessary speaking-tubes, with ivory or other mouth-pieces.]

No. 346.
Heating.

[HEATING.—Here provide the mode of heating by hot air, water, &c., whether high pressure or not, and specify engineer's name or apparatus; also provide for boiler, furnace, and all necessary fittings.
State number of coil boxes or chambers.
Or,
Specify the required kind of stove, "Gurney's" or other gill stove, flues, &c.] (See page 151.)

PLUMBERS' WORK.

No. 347.
Lantern top.

Cover the roof over lantern with 6-lb. lead, to fold round and under the edge moulding on top of fascia.

No. 348.
Ridges and Hips.

Cover the ridges and hips of roof or roofs with $\left\{ \begin{array}{c} \text{7-lb.} \\ \text{6-lb.} \end{array} \right\}$ lead $\left\{ \begin{array}{c} 16 \text{ to} \\ 20 \end{array} \right\}$ inches wide, securely fastened with lead-headed nails (and metal cramps if necessary), and closely dressed round ridge-roll and on to the slates.

No. 349.
Dormers.

Cover the (ridges and hips—or the tops) of dormer doors and windows with 6-lb. lead. (If *ridges*, say "as to roof,"—if *tops*, say "as top of lantern,"—or, if no lantern, describe the work similarly.)

No. 350.
Ditto.

Qy. Cover the sides of dormer doors and windows with 5-lb. lead.

| No. 351. Valleys. | Lay the valley gutters with 7-lb. lead, 16 inches wide, properly dressed under slates. |

| No. 352. Chimney gutters. | Lay guttering at back of (and, in superior work, at sides of) chimney-stacks, of $\begin{Bmatrix} 7\text{-lb.} \\ 6\text{-lb.} \end{Bmatrix}$ milled lead, to turn up against the stacks, and properly dressed under slates. |

| No. 353. Parapet gutters. | Lay the parapet gutters with $\begin{Bmatrix} 8\text{-lb.} \\ 7\text{-lb.} \end{Bmatrix}$ milled lead, to turn up $\begin{Bmatrix} 7 \\ 6 \\ 5 \end{Bmatrix}$ inches up the parapets, and to reach at least as high under slates. The gutter to have a medium width of   inches, and 2-inch drips every   feet, with a fall of not less than 2 inches in 10 feet. |

| No. 354. Flats. | Lay the flats with $\begin{Bmatrix} 8\text{-lb.} \\ 7\text{-lb.} \end{Bmatrix}$ lead, with all required roll-joints, to turn up against the walls, sill of lantern, &c., and (where there is no vertical boundary) dressed round and under the edge moulding of eaves fascia. The lead to have a fall of 2 inches in 10 feet, and drips if required. [Rolls should not exceed 27 inches apart.] |

| No. 355. | Roofs covered with lead, described, as flats. (See Zinc Worker.) |

| No. 356. Flashings. | Flashings of 4 or 5 lb. milled lead to be applied wherever the lead coverings of gutters, flats, or roofs turn up against vertical masonry or woodwork. Said flashings to be chased into walls at least 3 inches, and be dressed down over the lead turn-ups at least 4 inches. It is often advisable to carry the flashing quite through the parapet. |

| No. 357. Gutter cornices. | Line the gutters in eaves-wood cornices with 5-lb. lead (see Joiner) to fold over and under the front moulding, and turn up under slates two |

inches above the level of front moulding. Pieces of pipe with rose and boxing, to carry water from the gutter into the rain-water heads.

No. 358. Cisterns and troughs.
[Provide and fix earthenware, galvanised iron, slate, or enamelled cisterns, with overflow-pipes discharging outside house. [One cubic foot of cistern holds nearly $6\frac{1}{4}$ gallons of water; a gallon weighs 10 lbs. Cisterns for supply of the closets should be separate from that for domestic use.]

Or,

Line the cisterns with lead, the bottoms of 8-lb. cast lead, and the sides of 5-lb. milled ditto; and line the troughs conducting through the roofs with 6-lb. lead, or zinc.

[Specify charcoal or other cistern filter, as Atkins's or Ransome's filter.]

No. 359. Laying on water, &c.
Describe the pipes that may be necessary to conduct water into the cisterns from the town or Company's supply, the outer reservoir, or the force pump in ground-floor: also the waste-pipe required to prevent overflow, the dimensions thereof, the place into which it is to discharge, and any trap that may be required to prevent the ascent of effluvia from the drains below.

[Specify Tye and Andrew's or other approved sink traps to all sinks, &c.]

No. 360. Supply-pipes *from* cistern.
Describe any pipes required to conduct water to the pans of water-closets, or to any other parts of the building, as washing-places, baths, butlers' pantries, or housemaids' closets; stating such as are to have brass cocks, &c.

No. 361. Hopper Water-closet.
Provide and fix approved and complete valve hopper, or pan apparatus to water-closet, with all required brass pulls, levers, wires, cranks, copper ball-cocks, plugs, &c. Also soil-pipe, with ventilating outlet or pipe with trap to prevent the ascent of effluvia from the cesspit or drain. [Soil-pipes should be drawn or cast, not soldered. Iron or earthenware pipes are best.]

[Or,

Specify Jenning's earthenware valve closet in one piece, or Underhay's patent regulating, Tyler & Co's. or Lambert's apparatus.]

No. 362.
Linings of troughs, &c.

Line the washing trough in butler's pantry, or any other into which a pipe conducts from the cistern in roof; also bath (if of wood), &c., &c., &c., with lead of $\left\{\begin{array}{c}6\text{ or}\\5\end{array}\right\}$ lbs. to the foot: each trough, &c., &c., to have a brass plug and chain, and pipe to conduct therefrom into drains. Where water is brought down to supply jugs, pitchers, or pails, a shallow trough should be supplied to catch the droppings, with trapped pipe therefrom into drains or some movable vessel beneath.

No. 363.
Pump.

Provide and fix a (draw and) force-pump in the . . . . . where shown on plan, with -inch pipe thereto from the (well, tank, or reservoir), and pipe therefrom into the cistern in roof.

No. 364.
Sundries.

Describe any sheet lead that may be required in the joints of masonry, or for the covering of any parts thereof; also any necessary to cover the joinery, as the tops of wood cornices, parapets, &c. All nails used in Plumbers' work to be of copper. Outhouses and inferior buildings will require lead-work, as flashings, &c., occasionally, though there may be no lead on their ridges, hips, &c. Bell and clock turrets will be covered with lead in any circumstances.

[Specify all lavatories, urinals, &c., of approved manufacture, Mansergh's or other traps for wastepipes, &c.; charcoal or other ventilators. Provide Tyler and Sons' or other approved copper or galvanised tin bath, with copper pipes and $1\frac{1}{4}$-in. framing panelled with Honduras Mahogany top.]

No. 365.
General clause.

The lead-work to be laid and dressed down in the most careful manner, with as little soldering as may be, and with every regard for its expansion and contraction.

[Provide and fix complete all approved appliances.]

The work to be left by the Contractor perfect and complete, without any charge for the labour, solder, nails, holdfasts, joints, &c., which may be necessary to the efficient completion of the works herein generally described and partially particularized.

GLAZIERS' WORK.

No. 366. Glaze the windows of the .... with best patent plate glass (or the upper sashes thereof with sheet glass and the lower with plate); the windows of the .... with sheet glass (or the lower sashes thereof with flatted and the upper with best crown glass); the windows of the ... with best crown glass; the windows of the ... with glass of second quality; the windows of .... with third glass: the whole to be perfect in its kind, well puttied, and *left* perfect at the end of the works.

[Specify kind of plate glass, its thickness; also state quality or No. of sheet and crown glass. For horticultural buildings, Rendle's glazing is preferable to the common kind.]

PAINTERS' WORK.

No. 367. Paint the whole of the outside wood and iron work $\left\{ \begin{array}{c} \text{five or} \\ \text{four} \end{array} \right\}$ times in oil to finish a warm stone colour, and the inside do. do. (which it is usual to paint) $\left\{ \begin{array}{c} \text{four or} \\ \text{three} \end{array} \right\}$ times in oil to finish a .... colour.

[Paint an approved light olive, grey, or other tint, the stucco walls (if any).]

Paint also, in like manner (here state the extra painting, as) the treads and risers of stairs of .... up to the stair-carpet rings: the floors, from the skirting to the carpet line; continuing to particularise all such sashes, frames, shutters, soffits, backs, elbows, doors, jamb and soffit linings, frames, skirtings, panelled inclosures and linings, as are to be finished with any particular colour, or to be grained in imitation of some fancy wood and twice varnished. Finally, specify what *un*painted joinery (as real wainscot, &c.) is to be twice varnished or polished.

[Specify the "silicate," "petrifying," "oxide," or other approved paint for iron and external work.]

[Or,

No. 368. Stainer. Stain of an improved shade and twice varnish with the best copal all interior wood-work.

[Sometimes simply varnishing is preferred, in this case the panels and framework should be of picked deal or pitch pine of different shades.]

ZINC WORKER.

No. 369. [Here specify any flats, roof coverings, or other work to be executed in zinc. State kind of zinc, as that of the "Vieille Montagne Zinc Company," and the No. of gauge or thickness to be used; the kind of joint; ornamental hipping or cresting. No. 15, 24-oz. zinc should be used for flats; No. 13, 20-oz. for gutters. No. 14 gauge is better for good work.]

[*Note.*—In laying zinc particular care must be exercised in giving room for free expansion and contraction of metal, by adopting a roll over which the sheets may lap joint. No solder or nails should be used to connect the sheets; galvanised iron wall hooks to secure the flashings.]

[*Note.*—For buildings within the jurisdiction of the "Metropolitan Building Act" the young Architect must consult the rules, schedules, &c.]

CONDITIONS OF CONTRACT.

We append here an outline form for General Conditions for Building Contracts.

No. 1. The works are to be executed in the best and most workmanlike manner, and in accordance with the true and reasonable intent of the plans, drawings, and specifications taken together, which are to be signed by the architect and contractor, and in case of any discrepancy between the drawings and specifications the architect is to decide which is to be followed.

No. 2. No extra work is to be executed except by the express order of the architect in writing, and any variation made in carrying out the works is not to vitiate the contract, but all additions, omissions, or variations made in carrying out the works for which a price may not have been agreed upon, are to be measured and valued and certified for by the architect, and added to or deducted from the amount of contract as the case may be, according to a schedule of prices or at fair measure and value.

No. 3. Should any of the works be in the opinion of the architect executed with improper materials or workmanship, the contractor is when required by the architect forthwith to pull down and re-execute the same, and to substitute proper materials and workmanship, and in case of default of the contractor within a reasonable time, the architect is to have full power to employ other persons to re-execute the work, and the cost thereof is to be borne by the contractor.

No. 4. Any defects or other faults which may appear within... months from the completion of the building are, upon the direction of the architect, to be amended and made good by the contractor at his own cost; and in case of default, any cost incurred by the employer in so making good may be recovered by the employer from the contractor, the amount thereof in case of dispute to be settled as provided hereafter.

No. 5. The contractor is to insure the building against loss or damage by fire in an office to be approved, and the building is to be under the contractor's charge, who shall be responsible for, and make good, all damages occasioned by fire or otherwise, over which the contractor shall have control.

No. 6. The architect is to have at all times access to the works, which are to be under his own control, and he may require the contractor to dismiss any workman or workmen whom he may think incompetent or improper to be employed.

No. 7. The contractor to complete the whole of the works (excepting painting or papering, or other work, as the architect shall decide) within . . . . calendar months after the commencement, unless the works be delayed by reason of inclement weather, or causes not under control. In cases of default the contractor is to pay or allow the proprietor as by way of liquidated damages the sum of £ . . . . per week for every week during which the works shall remain incomplete.

No. 8. If the contractor shall become bankrupt, or make any arrangement for the benefit of his creditors, or shall delay the performance of his part of contract from whatever cause, the proprietor or his architect may give to the contractor or his assignee notice requiring the works to be proceeded with: and in default it shall be lawful for the proprietor or his architect to enter upon and take possession of the works, and to employ any other person or persons to carry on and complete the same. The costs and charges incurred in the said works are to be paid to the proprietor by the contractor, or may be set off against money due or to become due to the contractor.

No. 9. All work and materials intended to form part of the building are to be considered the property of the proprietor, and are not to be removed without consent of the architect.

No. 10. Any question or difference that may arise between the proprietor or architect and the contractor as to any additions made, or as to the meaning of the signed drawings and specification or any other matter or thing arising out of this contract, except as hereinbefore described, is to be referred to the arbitration and final decision of the architect, or in the event of his death and unwillingness to act, then of .... architect, or a Fellow of the Institute of British Architects, to be appointed on the request of either party.

No. 11.
(Here insert clause as to payments to be made on certificate of architect. Payments are usually made at the rate of 80 per cent. upon the value of works executed and materials delivered.)

MISCELLANEOUS HINTS AND CAUTIONS.

Union of new and old work.
In attaching any new work to a building, every allowance must be made for the sinking of the footings under pressure, and for the settlement of the masonry itself. Thus, while it is necessary that a vertical groove, or indent, be made in the old work, to receive a corresponding piece of the new, it is still more essential that a freedom for the downward motion of the latter 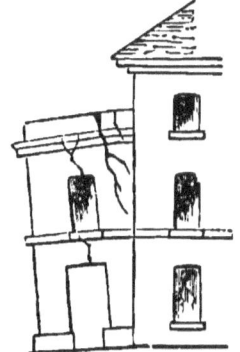 should be secured: otherwise, if it be tightly toothed and bonded into the old work, the result illustrated in the annexed sketch may be anticipated.

Union of ashlar facing with brick or rubble backing.
The same caution required in the latter case must be here equally observed. The *backing* (composed of small material and much mortar) will settle more than the *face;* and the latter will consequently bulge. This is easily remedied by computing, and allowing for, the difference of settlement; and by a due regard to the occasional bonding of the ashlar, so as to make the wall *one* substance, instead of *two* differently conditioned. The preceding  sketch illustrates the consequence of weight pressing upon unbonded ashlar and on yielding rubble.

P

[In thick walls concrete of Portland cement forms the best filling in bond, or through courses being introduced at intervals.]

Inverted arches.

Inverted arches must be used cautiously. Here is an instance, in which the points A and A were prevented by the inverted arch from sinking with the points B B, which latter sunk the more from the pressure of the arch C in the direction of the dotted lines. It is not uncommon for the young Architect to *affect* precautionary *science*, without a due consideration of the peculiar circumstances of his case.

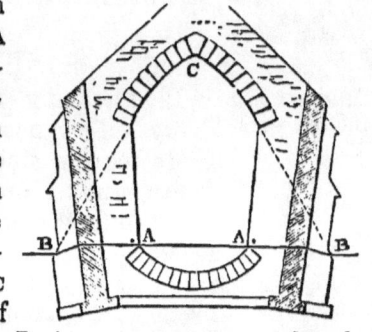

[Inverted arches should only be employed under a series of openings to equalise pressure of piers. Concrete foundations are best in other cases.]

Drainage, &c.

Always endeavour, if possible, to get your water-closet cesspit outside the building, so that it may be approached for cleansing without disturbing the interior. Be careful in the efficient use of dip-draps to prevent the ascent of rats from the outer sewer into the drains which are under the floors of the house. Rats are destructive in their operations, and if they die in the drain, prove, for a length of time, an unbearable nuisance. Drains may serve every purpose of carrying off soil and water; but the slightest opening in their upper part will allow the escape of effluvia into the space under the ground flooring, and thence into the rooms, unless that space be thoroughly ventilated with grated openings, allowing a thorough draught,—or, at least, a free ingress of fresh air and equal egress of foul. In the application of covered dry areas round the excavated basements of buildings, on no account omit their entire ventilation. If this be not attended to, the main walling, which they are intended to preserve from damp, may remain even more continually moist than if in immediate connection with the natural ground. Moisture frequently rises up the walling from below its foundation, and, exuding from the face of the masonry, remains confined, unless it evaporate and escape. Without means to this end, a covered area will be merely a receptacle for damp, and may keep the masonry continually wet, even when the ground outside is perfectly dry. Be especially cautious that the water from the rain-pipes of the roofs and flats be not conducted by them into the foundations.

[No drains should be placed under floors, and every trap should also be a ventilator.

Sewers and drains should have increased fall at their junctions and at all curves; the latter should be in the direction of the flow.]

Fire openings. It will save much subsequent trouble and disturbance of masonry, to be assured as to the size and character of the stoves, grates, ranges, &c., which the proprietor will employ. In the kitchen and cooking-rooms, especially, precautionary care should be taken in suiting the openings to the intended apparatus. Do not forget to be prepared for a smoke-jack, &c.

Sleeper or dwarf walls. In constructing these, do not omit the holes, &c., necessary for under-floor ventilation.

Paving. Be careful that the bottom, on which fine paving is laid, be dry and free from *staining* material. Common lime mortar is often injurious to pavements. Portland paving is especially liable to be disfigured by it.

[Wooden blocks burnetised, about $9'' \times 4\frac{1}{2}''$, and laid on concrete, make a good floor; hollow tiles or pots may be also used.]

Wrought stone-work.

In putting wrought stone-work together, *iron* is to be avoided as the certain cause of its subsequent destruction.

[Galvanised iron may be used, but not in great mass or long lengths. Space at ends should be allowed for free expansion.]

The stone cornices, architraves, and dressings of many a noble mansion have been brought into premature ruin by the contraction and expansion of iron under the effects of cold and heat. But there are careless Contractors who will allow their Corinthian capitals and fluted shafts to be ruined, even before the entablature surmounts them; and the young Architect will not, therefore, omit to insert a clause in his Specification (and to be peremptory in its enforcement), that all cut stone-work be securely preserved, during the progress of the building, with wood casing. It is surprising how grossly indifferent each class of artificers is to the work of the others. It is still more surprising to observe, how frequently they seem indifferent to the preservation of their own.

Slating.

Get rid of the Masons and Plasterers,—aye, and, as much as possible, of the Plumbers,—before your Slaters begin. The injury done to slating by the afterwork of chimney-tops, &c., is much to be dreaded. The cementitious "stopping" to a roof will not be efficiently done without close supervision : the ridge, hip, and valley courses will not be properly formed of large cut slates,—nor will every slate have its *two* nails, unless the Architect see to it.

Plastering.

Clear may be your Specification in forbidding salt sand, but, if your work be carried on in the vicinity of any estuary, the chances are (unless you be deemed cruelly strict) that the surface of your internal walls will vary with the weather, from damp to dry, like a sea-weed, and throw out salt in such abundance that you may sweep off a cellar-full.

[Clean road drift or coarse pit sand should only be used; fine gravel may be mixed.]

Beams, joists, and other timbers.— Lintels, bond, partitions.

It is the office of walls to carry beams, &c.; and that of beams to stay the walls from falling outwards or inwards: but it is the duty of Architects to see that the wood-work which supplants masonry does not weaken the latter; *i.e.* that the ends of timbers inserted into walls may not, by compression, or decay, leave the superincumbent masonry to loosen downwards. Thus, the beam A, though entering only a *portion* of the wall, presses upon the thorough-stone *e*, which 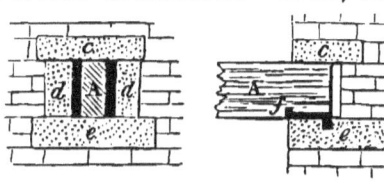 throws the weight upon the *whole* wall, and has, by means of an iron plate *f*, a hold to secure its perpendicularity. The cover-stone c presses on the surface of the timber to confirm its security: but should the timber rot, the cover-stone will not sink, because sustained by the side-stones *d d*. To *prevent* rot, the backing and side-stones are left free of the timber, so that air may traverse round it. The habit of placing the ends of beams on a templplate, as G, is bad. The only justification of the employment of wood, so built into the walls, is when it forms a continuous plate, that it may act as a bond to preserve the perfect horizontal level of joists, which, however, should extend a little beyond the plate, so

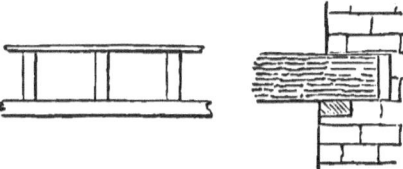

as to have a bearing also on the solid of the wall.

[A good plan is to form brick or stone corbelling to receive the plate and joists.]

Beams, joists, and other timbers.— Lintels, bond, partitions.

Careful inspection will then so manage the construction of the wall in this part, as to leave it but little weakened by the air hollows required for the plate and joists; unless, indeed, it be very thin,—as only one brick, for instance,—when no law of common sense can justify the use of continuous bond. Where joists uninterruptedly cross a thin wall, which is to support another story of masonry, let there only be one plate, thin, and on its edge, in the centre of the wall, so that at least a brick on edge may be placed on each side of it, to fill up the intervals between the joists, and give solid support to the superincumbent masonry. On no account let the upper part of the wall be separated from the lower by a mere layer of perishable wood, or supported by a range of joists on their edge. It has often occurred to us that iron hooping should be more used than it is as the internal bonding of walls. At the same time, it must be remembered, that bond timbering is necessary, at intervals, to receive the nails of the battening. When, however, the wall is thin, it may be imperative to avoid its use, employing old oak bats for that purpose.

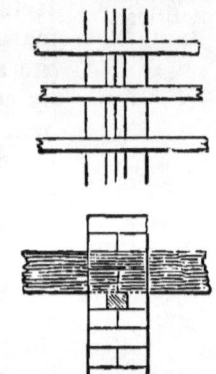

[We think it best in the case of a thin wall to have brickwork corbelled out on both faces to receive ends of joists—joists running through walls conduct sound, also fire, &c.]

In short, let it be the care of the young Architect, so to contrive the union of his masonry and carpentry, as that the entire removal of the latter may leave the former secure in its own strength. In the use of *lintels* especially, he should be cautious. They are useful as bonds to unite the tops of piers, and as means for the fixing of the joinery; but they ought never to be trusted to as a lasting support of masonry,—that support being always really afforded by the relieving segment

HINTS TO YOUNG ARCHITECTS.

Beams, joists, and other timbers.— Lintels, bond, partitions.

arch above the lintel. We are aware, that a bressummer may be termed a large lintel; and that, here, at least, the support of the masonry is truly intended. The use of the bressummer, in shop-front openings, is an evil necessity to which we must often submit; and all that an Architect can do, is to make the best of a bad job, by *wrought*-iron trussing, which will at least give adequate *strength*, though it may not insure permanent durability. If *time* spare it, *fire* may destroy it; and the latter evil is not to be met even by iron, which, if wrought, will bend,—if cast, will crack, with heat. Let the arch, then, or some modification of it, be always used—if possible.

[A self-tied arch of brick, tied by an iron flitch plate in centre, is recommended, or tie rods and cast-iron skewbacks may be used.]

Partitions of wood should not be left to the sagacity of the Carpenter. Under all circumstances where they have to support themselves over voids, or to bear, or participate in the bearing of, a pressure from above, they should be considered by the Architect in his Specification, and carefully studied in making the working drawings. It is not enough, merely to say, that "they are to be trussed so as to prevent any injury to ceilings by their own pressure, or that of the roof above them;"—marginal sketches should be made, showing the disposition of the skeleton framing, with whatever iron-work is necessary to its security.

See, for instance, what a Carpenter may do, unless well directed: a roof c, bearing partly on the partition A, when it should have borne only on the walls; and, instead of distressing the partition, should have rather held it suspended; the partition A bearing down with its own weight, and that of the roof, on the

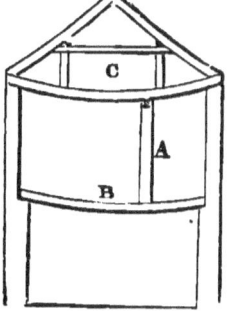

Beams, joists, and other timbers.—Lintels, bond, partitions.

floor B, instead of being so truss-framed in its length as to leave the floor unconscious of its existence. We presume no ignorance in the young Architect as to the manner of doing these things; and only call on him not to suppose they are so obvious as to be done without his guidance.

[All partitions should be trussed by braces springing from the supports, iron king-bolts being used to prevent centre portion sagging.]

In the framing of roofs, give a maximum strength to the purlins: the undulating surface of a weakly purlined roof will soon proclaim its defect in this particular.  The position of the principals should not be observable from without.

[All principals should be self-sustaining and rest on walls.]

Floors; simple and framed, &c.

For permanent and uniform strength, there is no floor so good as one composed of simple joists, stiffened by cross bonding: but, in very large rooms, there is more economy in the compound floor of binders and joists, or of joists, binders, and girders. There may be particular reasons for girders, &c.; as, when the weight of the floor has to be thrown upon piers, and not on a continuous wall of uniform strength: but the usual motive to the use of the compound floor, in rooms which exceed 18 or 20 feet in width, is a legitimate economy of materials. It is only necessary to caution the young practitioner on the necessity of considering, that girders have to perform the duty of cross walls; that they should be trussed to prevent their "sagging," even with their own weight; that their scantling should allow for the weakening effect of the cuttings made into their substance to receive the timbers they support; that their trusses should be wholly of *iron* (and not partially of oak, for the Author

HINTS TO YOUNG ARCHITECTS. 321

Floors; simple & framed, &c.

has seen the bad effects of the shrinkage of oak struts); and, especially, that the end of each girder, instead of being notched on perishable templates of wood, and close surrounded with mortar and masonry, should be housed in a cavity (as we have already described) with an iron holding plate; or inserted into a cast-iron boxing, notched into a thorough-stone, leaving a space (however small) for air to circulate about it, and prevent rot. The failure of a girder involves the failure of all the rest of the floor; and, though *all* timbers inserted in masonry should have a more careful regard to their preservation from decay than it is usual to bestow, it will be readily admitted, that too much care cannot be given to those leading bearing timbers, without the permanent duration of which the durability of the large remainder is of no avail.

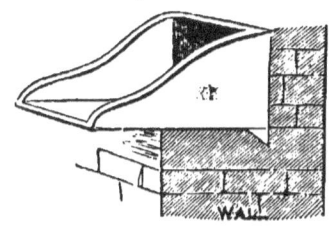

[Rolled iron joists with cross joists or hollow rebated tiles or brick arches are best for large floors. (See Fireproof Construction.)]

Roofs.

The same remarks, applying to the extremities of girders, apply also to tie-beams.

Ceilings.

To procure a good ceiling in single-joist floors, it is necessary there should be ceiling joists crossing below the others: and it is a question whether the ceiling joists, under double-framed floors, instead of being chase-mortised *into* the binders, should not be in unbroken lengths nailed *under* the binders. Where the ceiling joists (as under roofs) are likely to be trodden upon, they must be well secured.

Sound boarding.

Always consider whether the occupants of any particular room will be annoyed by the noises of the rooms below or above. Sound boarding and

P 3

pugging considerably increase the weight of the floor, the scantling of whose timbers should therefore be thought upon. Water-closet partitions should be well pugged.

[Pugged floors with mortar and hay should always be adopted.]

Mice in partitions and skirtings.
The space behind the skirtings is often a thoroughfare for mice, which also contrive to travel from floor to floor in the hollows of the quarter-partitions, and become in several ways a great nuisance. Plaster or wood stopping is not always so efficacious as the use of broken glass in those secret passages which they are prone to frequent.

Coverings to gutters, cisterns, &c.
The liability of gutters and cisterns to become choked with snow, or infested with leaves, &c., renders it advisable to protect them with a boarded covering, which may preserve the under current of water from receiving what may speedily produce a chokage or overflow.

[It is a good plan to have two cisterns, one for drinking purposes, and the other for flushing drains. Cisterns should be of slate or earthenware glazed, or of galvanised iron.]

Iron columns, beams, &c.
On this most important subject we say but little, that we may signify the more. Here, the young Architect should not move a step without carefully consulting the experienced knowledge of the Engineer. Tredgold's "Practical Essay on the Strength of Cast Iron," Hodgkinson's and Fairbairn's works, should be well studied whenever necessity compels the support of heavy and loaded superstructures by iron columns and beams. A careful computation of the weight of the mere building, added to that of its possible burthen, with allowance for theoretical fallacy, and a due estimate of the increased strength of the hollow pillar, as compared with a solid one having the same amount of metal, must be made,

Iron columns, examined, and re-examined, before the Specifica-
Beams, &c. tion be issued.

[Under the bearing parts of all columns, sheet lead or a packing of iron and Portland cement grouting should be laid.]

GENERAL INDEX.

ABUTMENTS, stability of, 114, 121; thickness of, 122.
Additions and alterations, 94.
Æsthetic elements, 179, 189.
Agreements, drawing up, 95; form of, 96.
Air, heated, 155; circulation of, 159; quantity necessary to respiration, 161.
Alison on Beauty, 190, 193.
Angle, limiting, of resistance, 113; of repose, 116, 117; of rupture, 122.
Angular perspective, 63.
Apparent and real size, 75.
Arch, equilibrium of, 118; Gwilt on, 29; angle of rupture in, 122; inverted, 300; theory of the, 118; line of pressure in, 120; arched ribs, 130 (see Specification, No. 56 et seq.).
Architects, duties of, 79; education of young, 32; inclination of those intended for, 2; position of, 97; responsibility of, 31.
Architecture, its claims as an art, 32; abstract principles of, 180; principles of design in, 179; laws of design in, 181; mechanical principles of, 181; taste in, 173; style in, 177.

Architectural studies, 37; abroad, 41; tourists, 49.
Architrave (see Specification, No. 75, &c.).
Archivolts (see Specification, No. 80, &c.).
Art, affectation in, 177; classification of, 190; expression in, 190; knowledge of, 45, 51; qualities necessary to, 191.
Articleship, hints to pupils during, 37.
Ashlar, 227 (see Specification, No. 94).
Asphalte for damp proof course, 216; for paving, 219 (see Specification).
Axis, neutral, 137.

BAIN on Beauty in Art, 191.
Balance and stability of structures, 110.
Barge board (see Specification, No. 226, &c.).
Baths (see Specification, Nos. 138, 362).
Beams, strength of, 123, 142; position of, 125; stiffness of, 140; timber and iron, 136; cast iron, 145; wrought iron, 146; rolled iron, 147; flitch, 143.

Beauty, definition and hypotheses of, 190; Bain on, 191; of form, 190.
Bond (see Specification, No. 12); hoop iron, 34, 185.
Brick and stone arches, 118; brickwork, 216; paving, 219 (see Specification, bricknogging, No. 24).
Bressummers, 143 (see Specification, No. 187).
Building committees, 93; sanitary conditions of, 171; builder's work, measurement of, 29.
Buttresses, 112 (see Specification, No. 104).

CARPENTRY, 133 (see Specification).
Casements, 77 (see Specification, No. 281 *et seq.*).
Cast iron, 145; beams, 145; columns, 138; design in, 187; cast metal-work, &c., 189.
Cast shadows, 21.
Ceilings, 307 (see Specification, No. 163, &c.).
Cement, 197; selenitic, 198; cement work (see Specification).
Centre of gravity, 108; of pressure, 116.
Ceramic materials, 188.
Chimneys, 68 (see Specification).
Cistern, 201; connection with soil pipe to be avoided, 171; waste pipes from, 71; covered, 172.
Classic architecture, 174; stonework (see Specification, No. 64, &c.).
Colour, 195.
Columns, appearance of, 24; stone and brick, 138; cast iron, 139; hollow, 139; resistance of, 138; steel, 148.

Combination of materials, 199.
Composition, connection in, 69.
Concrete walls, 200; foundations, 215 (see Specification); value of, as fire-resisting, 200.
Construction, principles of, 99, 197.
Contracts, drawing up, 95, 310.
Contrast, 192.
Cornices, value of, 76; stone (see Specification, No. 89, &c.).
Cramps, 198 (see Specification, Nos. 12, 126).
Crushing strength of columns, 139.
Curve, circular, 13; to draw, 14; curvilinear forms, 192, 193.

DAMP-PROOF course (see Specification, No. 13).
Dams, pressure of water against, 118.
Details, 81.
Dining-room, 88.
Domes, stability of, 123.
Doors, double, 157; sliding (see Specification, No. 251).
Doorways, position of, 84.
Drains, 72; main brick (see Specification, No. 44); rubble (No. 43); stoneware, 38.
Drawing, 4; methods of, 9; of forms, 13.
Dressings (see Specification, No. 75, &c.).

EARTH, angle of repose, 117; earthwork (No. 3, &c.).
Eaves (see Specification, Nos. 207, 225).
Egg-shaped sewers (see Specification, No. 44).
Elasticity, modulus of, 137.
English Gothic, 46, 176.
Entasis of columns, &c., 193.

GENERAL INDEX. 327

Expansion of materials, 199.
Expression in architecture, 194.

FIREPLACES, position of, 84; open, 151.
Fittings, stable (see Specification, Nos. 140, 309, 340.)
Flashings (see Specification, No. 356).
Flats (see Specification, Nos. 197, 354).
Floors, 142; weights of, 135, fireproof, 306 (see Specification, Nos. 192, 194).
Forces, equilibrium of, 99; moments of, 104.
Forms, delineation of, 12; beauty of, 190.
France, studies in, 52; French language, 31; French casements, No. 285.
Frieze (see Specification, No. 119).

GABLES (see Specification, No. 109).
Generalisation of facts necessary, 34.
Geometry, useful study of, 2; works on, 3.
Girders, cast iron, 145; wrought iron, 146; box, 147; lattice, 147.
Gothic compared with Greek, 183; English, 176; continental, 176; Italian, 46, 151, 178; roof (see Specification, No. 203).
Gradation, 192.
Gravity, centre of, 108; of various figures, 109.
Greek, 174; static element in, 181; character of, 183.
Gutter cornice (see Specification, No. 223).

HARMONY, 191; harmonic proportions, 191.

Hearths (see Specification, No. 129).
Heating, 151; conservation of heat, 156 (see Specification, No. 346).
Height, apparent, obtained, 194.
Hints on perspective, 21; to tourists, 49; on form delineation, 12; miscellaneous, 299.
Hogarth's principles of beauty, 190.
Hollow columns, 139; ditto walls, 156, 198 (see Specification, No. 12).
Hoop iron, 200 (see Specification, No. 34).
Hydraulic lime, 197.

INERTIA, moment of, 138.
Inverted arches, 300 (see Specification, No. 16).
Iron, cast, 145; cast beams, 145; ditto columns, 139; construction, 144; wrought, 146 (see Specification, No. 322); coverings, 136; galvanized, 302 (see Specification, " Ironwork "); fittings of stables (No. 340); expansion of, 199; iron and metal work, 283; gutters (No. 327).
Isometric projection, 25.
Italian studies, 31, 49; Gothic, 46, 178.

JOINERY (see Specification).
Joints in carpentry, 133 (see Specification).
Joists, depth of, 142; iron, rolled, 147 (see Specification, Nos. 190, 209, 322).
Junction of house and offices, 67; ventilating, 165; drain, 72 (see Specification).

GENERAL INDEX.

LATIN, 30.
Lattice girders, 147.
Leadwork (see Specification).
Lever, principle of the, 106.
Light, 78; light and shadows, 16.
Lime, 197 (see Specification).
Limiting angle of resistance, 113.
Loads on floors and roofs, 136.
Lodges, 92.

MAGNITUDE, centre of, 108.
Masonry (see Specification).
Materials, their functions, 184; strength of, 136.
Mathematical studies, 2.
Measurement of builder's work, 29.
Mechanical principles, 99.
Do. of architectural design, 181.
Metal-work, 189.
Methods of study, 11; drawing, 9.
Modes of failure of walls, 113.
Modulus of elasticity, 137.
Moments of forces, 104; of inertia, 138.
Mortar, 197.

NATURAL inclinations, 1.
Natural slopes, 117.

OCULAR correction, 193; impressions, 74.
Optical correction, 193.
Ornament, proper use of, 82.
Oval sewer, method of describing, 211 (see Specification).

PAINTING (see Specification).
Parallelogram of forces, 99.
Parapets (see Specification, Nos. 90, 100).
Parian cement (see Specification).
Parthenon, its proportions, 174; static principle in, 184.
Partitions, 305 (see Specification).
Papier-mâché, design in.

Paving (see Specification, Nos. 25, 131, &c.).
Pediment (see Specification, No. 108).
Perspective, eye, 14; projection, 16; hints on, 21; isometric, 25.
Pillars, strength of, 139.
Pipes, drain, 201; supply, 201 (see Specification, No. 38, &c.).
Planning, hints on, 58; reference of, to elevation, 60.
Plastering, 302.
Plastic substances, 188.
Plate girders.
Plinth (see Specification, Nos. 87, 97).
Plugs (see Specification, 126).
Plumbing, 71 (see Specification).
Pointed arch, 121; domes, 123.
Polygons, to draw, 13.
Portico (see Specification, Nos. 111, 112).
Portland cement, 198 (see Specification).
Precision of habits, 6.
Pressure, centre of, 116; of earth, 116; wind, 116; water, 116, 118.
Principles of study, 33; of construction, 99; mechanical, 181.
Projection, perspective, 16; isometric, 25.
Proportion in design, 191; of rooms.
Pump (see Specification, No. 363).
Pupils, 7; hints during pupilage, 37.
Purlins, scantlings of, 143.

QUANTITIES, 95.
Quarter partitions, No. 188.
Queen truss, 129.
Queen slating (see Specification, No. 154).

GENERAL INDEX.

Quoins (see Specification, No. 92).

RAFTERS, 143 (see Specification, No. 199, &c.).
Resistance, limiting angle of, 113; to compression, 138; to cross-strains, 140; line of, 115.
Responsibility, 31.
Retaining walls, 116.
Reveals (No. 55).
Rolled iron joists, 147.
Roman architecture, 47, 49, 174.
Roofs, trusses, 129; loads on, 136; coverings of, 136 (see Specification, No. 199, &c.); open, 203, 355, &c.; Gothic roof, No. 203; coverings, 200.
Rubble masonry (see Specification, No. 50, &c.).
Rusticated work (No. 93).

SANITARY construction, 151; conditions, 171.
Sash windows (No. 256, &c.).
Scagliola work (No. 171).
Scantlings of timber, rules for, 141 (see Specification, No. 181 et seq.).
School studies, 1, 7.
Selenitic cement, 108.
Sewerage, 162.
Shadows, 16.
Sinks (see Specification, No. 137).
Sketching, 15; advantages of, 61; abroad, 45, 50.
Skylights (No. 217, &c).
Slating (No. 143 et seq.)
Solids and voids, 66.
Sound boarding (No. 195).
Spire, entasis in, 193.
Stability of structures, 110; of beams, 123.
Stables, 88 (see Specification, Nos. 309, 339, &c.).

Stairs (see Specification, No. 290, &c.).
Statical design, 181.
Steel, 148.
Steps, 216 (see Specification).
Stiffness of beams, 140.
Stone, 198; stonework (No. 64 et seq.).
Strength of materials, 136, 142.
Studies, method of, 11; mathematical, 2; principles of study, 33; in France, 52.
Style, arrangement should suggest, 60.

TASTE, architectural, 173.
Technical studies, 28.
Tension, 138.
Timber, 136, 186.
Tiles (No. 156, &c.); encaustic, (No. 26).
Tourists, hints to, 49.
Tracery, to draw, 13.
Traps, drain, 71, 165.
Travel, 42; benefits of, 43; time occupied in, 48; hints on, 49.
Triangle of forces, 101; centre of gravity of, 109; triangular plans, 65; frame, 127.
Trusses, 126.

UNIFORMITY in design, 192.

VALLEYS (see Specification, No. 351).
Vaults, 112 (see Arch) (see Specification, No. 28, &c.).
Venetian windows (No. 278).
Ventilation, 157; ventilating traps, 165 (see Specification); under floors (No. 48); of water-closets (No. 219).
Verandah (No. 341).

WALLS, stability of, 111; retaining, 116; hollow, 156 (see Specification); sleeper, 23, &c.
Water-closets, position of, 70; fittings to, 201 (see Specification, Nos. 306, 361); ventilation of (see Specification).
Water-pipes (No. 360).
Water-tank, 47.
Well-digging (see Specification, No. 8).
Windows, 76; proportion of, 78; sash (see Specification, No. 266, &c.); bay, 276; dressings, 79, 83, &c.
Woodwork 187 (see Specification, under "Carpenter and Joiner's" work).
Wrought iron, 145; ditto beams, 147; plate girders, 147.

ZINC, 136.
Zinc worker (see Specification).

THE END.

PRINTED BY J. S. VIRTUE AND CO., LIMITED, CITY ROAD, LONDON.

www.ingramcontent.com/pod-product-compliance
Lightning Source LLC
Chambersburg PA
CBHW032050220426
43664CB00008B/936